HOWARD W. SAMS

Guide to HDTV Systems

HOWARD W. SAMS

Guide to HDTV Systems

By

Conrad Persson

PROMPT®

PUBLICATIONS

International Standard Book Number: 0-7906-1166-X
Library of Congress Catalog Card Number: 98-066675

Acquisitions Editor: Loretta Yates
Editor: Pat Brady
Assistant Editors: J. B. Hall
Typesetting: Pat Brady
Indexing: Pat Brady
Proofreader: Patika Bush
Cover Design: Christy Pierce
Graphics Conversion: Brian Rees
Illustrations: Courtesy Advanced Television Systems Committee

PRINTED IN THE UNITED STATES OF AMERICA

9 8 7 6 5 4 3 2 1

3 9082 07901 4604

Contents

INTRODUCTION

What is HDTV (High-definition television)? For one thing, it's digital. That means that the picture and sound information are converted from their normal analog condition at the transmitting end, into their digital representation. That digital information (picture and sound) is imposed onto a carrier signal and sent as a stream of bits (bit stream) to the receiver.

Because it's digital, the picture displayed by the receiving device will be a faithful reproduction of the picture that was sent by the transmitting device. The displayed HDTV picture will not show any of those imperfections that we sometimes see in an analog TV broadcast: ghosts, snow, sparklies. The HDTV picture will either be perfect, or it will not be received at all.

High-definition also means...well, high-definition. The broadcast standard currently in use in the United States is the standard that was developed decades ago by the National Television Standards Committee (NTSC). The NTSC signal is also going to be 16X9, but there's more; and the sound.

As you read this book, you will also become aware of another important fact of HDTV. The language describing HDTV actually sounds more like the language describing computer/digital communications than the language of television. There's a reason for that. The transmitter/receiver link in HDTV is really a digital communications path. The HDTV signal is made up of streams of bits. As you will see in reading this book, the video and the audio are digitized separately, arranged in streams, separated into packets, processed for transmission, and then sent out over the airwaves. At the receiver, the reverse process occurs.

Just as with computer communications, you will read about "bits," "bytes," "protocols," various "layers," "data," and "metadata," and more.

A BLOCK DIAGRAM OF THE SYSTEM

The HDTV transmission system is shown in block diagram form in *Figure 1-1*. At the transmission side, the HDTV system is conceived of as consisting of three subsystems:

1. Source Coding and Compression

2. Service Multiplex and Transport

3. RF/Transmission

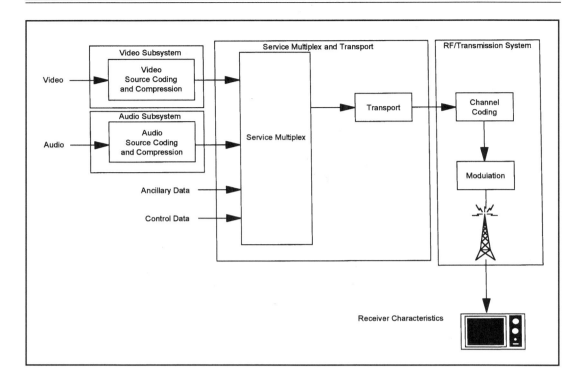

Figure I-1.

Source Coding and Compression

In the first subsystem, "source coding and compression," all of the information to be broadcast is converted from analog to digital form and the digital information is compressed. This includes the video information, audio information, and "ancillary" digital data. The "ancillary data" includes control data, conditional access control data, and data that is associated with the audio and video program data, such as closed-captioning. Ancillary data can also include independent program services.

The coder minimizes the number of bits required to represent the audio and video information. The HDTV system uses one scheme to code and compress the video, and another to code and compress the audio. For the video, HDTV uses a scheme developed by the Motion Picture Experts Group (MPEG), called MPEG-2. For the audio portion of the signal, HDTV uses the Digital Audio Compression (AC-3) Standard.

Service Multiplex and Transport

In the HDTV subsystem referred to as "service multiplex and transport," each digital data stream—video, audio, and ancillary data—is divided into "packets" of information. Each packet is given a unique identification, then the audio data stream packets, video data stream packets, and ancillary data stream packets are multiplexed into a single data stream.

When this HDTV transport system was designed, the designers were mindful of more than just terrestrial broadcast of the signal. They were designing it so that the signal could be compatible with broadcast, cable, satellite, recording and computer interfaces.

RF/Transmission

The RF/Transmission subsystem is the segment of the system in which channel coding and modulation take place. When the HDTV signal is transmitted, whether via terrestrial broadcast, cable, satellite, or whatever, it is subject to impairment and degradation, so that when it reaches the receiver, it may not be exactly what was transmitted. To compensate for these problems, the channel coder portion of the RF/Transmission system adds additional information to the data bitstream that can be used by the receiver to reconstruct the data from the received signal.

The modulation section of the RF/Transmission, also known as the physical layer, uses the digital datastream to modulate the transmitted signal. The modulation subsystem can operate in two modes: a terrestrial broadcast mode (8 VSB), and a high data rate mode (16 VSB). VSB stands for "vestigial sideband."

As you read this book, keep in mind that the ATSC Standard describes a very complex system, and the technology used in the system is constantly being improved and updated. Therefore, the ATSC Standard, and thus the products manufactured to that standard, and the descriptions of the system in this book, are subject to change. For more information about the ATSC, check the Web site at: http://www.atsc.org.

1

VIDEO
COMPRESSION

Creation of a high-definition television picture, along with the associated high-quality digital audio, takes a lot of information. In fact, in order to send an HDTV signal without first compressing it would require that the data be sent at a rate of 1 Gbps (one gigabit, or 1 billion bits per second). A standard 6 MHz television channel such as the ones that have been used for decades to transmit NTSC television signals are capable of reliably transmitting about 20 Mbps (20 megabits, or 20,000,000 bits per second). The math here is pretty easy. If we don't compress the data stream before we send it, we're going to have to use a portion of the available frequency spectrum that's the equivalent or 50 NTSC channels. Not a good idea, according to the FCC.

In order to send all that information within a 6 MHz bandwidth, we're going to have to cram that huge amount of data into a single narrow little channel, we're going to have to smoosh it into a package that's 1/50 its original size.

COMPRESSING THE VIDEO SIGNAL

The method used in HDTV to achieve data compression of the video bit stream is based on mathematical algorithms that are so abstruse that they'd really only be understood by a relative handful of pretty bright engineers and mathematicians, so we won't even try to go into that in this book. What we'll try to do instead is to give an idea of what's going on that will be understandable for most of us ordinary people.

For starters, keep in mind what a motion picture is all about, whether it's shown on the big screen down at the local multicinema complex, or on the TV screen in your own living room. The motion picture consists of a series of still frames. The frames are displayed at a rate of about 30 per second in the case of television pictures. Because of the rapidity with which the frames follow each other, the eye/brain system is fooled into perceiving this as continuous smooth motion.

If you'll sit and watch a movie or TV program attentively some time, you'll get an idea of how the information required to represent that sequence of images might be compressed. For example, let's say that you're watching a tranquil scene on a desert island, maybe on PBS. Looking out to sea, you can see the azure sky, the blue of the sky almost exactly the same throughout. The sand on the beach is, well, sand color, and again, the color and brightness vary little throughout. The only thing going on is the gentle motion of the waves.

Temporal Compression

One kind of compression that can be achieved in transmitting that scene is by transmitting the complete scene only in one, or a few frames. Now as long as the camera remains in one place, subsequent frames could contain only as much information as necessary to update the position of the waves as they wash upon the shore. This type of compression is known as "temporal" compression: that is, it's compression based on those portions of the scene that change, or don't change, over time, and therefore much of the scene doesn't change from frame to frame.

Of course, if the scene happens to be a fast break during a basketball game, or a series of car chases and explosions, temporal compression is not going to be as easily achieved.

Spatial Compression

Then there's "spatial" compression. In that tranquil desert island scene we mentioned earlier, the color and brightness of the sky and the sand are very consistent throughout. It is not necessary, then, to transmit the color and brightness information for every tiny bit of each of those areas. We could instead send the color and brightness information for a portion of the sky and then tell the receiving set which other areas are sky or sand, and to make the rest of them the same color and brightness. That's an oversimplification, and might even be a little misleading, but it does give the idea of how the MPEG standard achieves compression.

A Slightly More Technical Explanation of Compression

The MPEG-2 specification is organized into a system of "profiles" and "levels" so that it will work for not only terrestrial broadcast, but for cable, satellite, and even computers, provided that they use the appropriate hardware and software. The Digital Television Standard portion is based on the MPEG "Main Profile." Now here's where it gets complicated. In order to achieve data compression, as defined by the MPEG main profile, there are three types of frames for prediction: I-frames, P-frames and B-frames (more about these frames later).

Here's a definition of the term "frame," so you won't have to turn to the *Glossary* section. A frame contains lines of spatial information of a video signal. For progressive video, these lines contain samples starting from one time instant and continuing through successive lines to the bottom of the frame. For interlaced video, a frame consists of two fields, a top field and a bottom field. One of these fields will commence one field later than the other.

In addition, the MPEG main profile also includes an organization of luminance (brightness) and chrominance (color) samples within the frame.

AN OVERVIEW OF VIDEO COMPRESSION

The analog video source signal is fed to the video compression system. The output of the system is a compressed digital signal that contains information that can be decoded to produce a sequence of images that is very close to the original video signal. The goal, as stated by the Advanced Television System Committee (ATSC) is for the reconstructed signal to be imperceptibly different from the original, for most viewers, for most of the images, for most of the time.

Figure 1-1 is a block diagram of video coding, showing signal flow. The analog signals that are input to the system are digitized and then sent to the encoder for compression. The compressed data is then transmitted over a communications channel. Once the signal has been received by the consumer's TV set, it is decompressed in the decoder. The transmission medium may, of course, have caused corruption of the signal. In this case, the decoder would use the additional information that was encoded at the transmitting end to reconstruct the image for display.

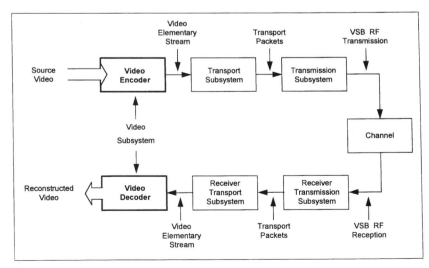

Figure 1-1

Video Preprocessing

The analog video signals that are presented to the preprocessor are the red (R), green (G), and blue (B) signals. In video preprocessing, these analog signals are converted to digital samples in a form that can be used in the compression process that is to follow.

Formats for Video Compression

There are four formats allowed in the Digital Television Standard, as shown in *Table 1.1*. As will be discussed in the receiver portion of this book, HDTV receivers are expected to be able to accept any of these formats, convert them, and display the video content in the receiver's "native" display format.

Vertical lines	Pixels	Aspect ratio	Picture rate
1080	1920	16:9	60I, 30P, 24P
720	1280	16:9	60P, 30P, 24P
480	704	16:9 and 4:3	60P, 60I, 30P, 24P
480	640	4:3	60P, 60I, 30P, 24P

Table 1.1 Compression Formats

In the table, "vertical lines" refers to the number of *active* lines in the picture. The term "pixels" refers to the number of pixels that make up an active line. The term "aspect ratio" describes the ratio of the picture width to picture height.

The term "picture rate" deserves a little extra bit of explanation. In the designations for picture rate, the numeric describes the number of frames or fields per second. Note that each numeric is followed by either the letter "I," or "P." The letter I indicates interlaced scanning, while the letter P indicates progressive scanning. In the case of interlaced scanning, each picture "frame" is made up of two picture "fields," which are interlaced to form a complete picture.

Representation of Picture data

In digital television, the information that describes an image is in digital form. This allows the data to be processed digitally, not unlike that used in a computer. Digitization is the process of sampling the analog television signals and their components, and using a digital code to represent each sample.

Film Mode

Video encoders have the capability to recognize that when a large percentage of the pixels don't change from frame to frame in the image sequence that the input was film, with a rate of less than 60 frames per second (fps).

Motion pictures, for example, are displayed at 24 fps. When the source of the video information for broadcast is a motion picture film played at 24 fps that is sent at 60 Hz using what is called a 3:2 pull-down operation, the processor may recognize that the input was 3 nearly identical pictures followed by 2 nearly identical pictures. In this case, it would only have to encode the unique 24 images per second that comprised the original material. When the encoder detects that the input is 24 fps film, the input signal is converted back to a progressively scanned sequence of 24 frames per second before the data is compressed.

The benefit of this film mode is that it avoids sending the same frame many times, which, of course, results in a much better compression ratio. When this mode is invoked, the encoder includes this information in the ancillary data that is included with the bit stream, which is then used by the receiving television set in correctly decoding the signal.

In a similar manner, if the processor receives 30 fps film material sent at 60 Hz, it might recognize that it was receiving two almost identical pictures followed by two almost identical pictures. When the encoder recognizes this fact, the input signal would be converted back to a progressively scanned sequence of 30 fps And again, this information would be conveyed via the ancillary data stream to the receiving television set so that it could properly decode the signal.

HANDLING THE COLOR INFORMATION

The video signal that is the electronic representation of the original scene is sent to a matrixing system that converts the original red, green, and blue (R, G, and B) information to luminance (brightness, represented by the symbol Y) and color value components, called chrominance, and represented by the two symbols Cb and Cr. At the receiving set, the Y, Cb, and Cr values are combined in such a way as to recover the original R, G, and B information.

Luminance vs. Chrominance Samples

It's important to keep in mind that the picture information consists of the luminance, or brightness, information, and the information that is needed to describe the chrominance, or combination of Cb and Cr that combine to create the colors in the original scene. This means that coding of the input video signal involves sampling all of these values.

The original RGB components tend to vary in step with each other, and so are said to be "highly correlated." The Y, Cb, and Cr signals are not as correlated, which makes it easier to code them efficiently.

The differences in the way in which humans perceive luminance and chrominance allow the coding process to be performed more efficiently. Given a point in an image described by Y, Cb and Cr values, most of the high frequencies occur in the Y component. Because of the nature of the human visual system, it is more sensitive to high frequencies in the luminance component than to high frequencies in the chrominance components.

To take advantage of these characteristics in making the coding process more efficient, the chrominance samples are sent through a low-pass filter, which removes some of the higher frequencies, and sub-sampled at half the rate at which the luminance signal is sampled, along both the horizontal and vertical dimensions. This results in chrominance components that have one-fourth the spatial resolution of the luminance component.

Pixels

The sequence in which the analog video signals are sampled corresponds to the manner in which the video picture is displayed by the scanning raster: from left to right within a frame, and in lines from top to bottom. The group of samples that represents a single frame, or, in the case of interlaced scanning, a single field, is treated as a unit, as if they existed at a single instant in time. The individual samples of image data are called picture elements. The term "picture element" is frequently abbreviated to "pixel," or "pel." One frame of sampled video, or, in the case of interlaced scanning, one field, can be considered to be a rectangular array of pixels.

In the case in which the ratio of active pixels in each line is the same as the ratio of the picture width to picture height, or aspect ratio, which in HDTV is 16:9, the image is said to have "square" pixels. Don't get the idea that this means that the pixels are square. They're not. In fact, the ideal for a pixel might be to have zero area. The term "square" refers to the *spacing* of samples, not to shape of the pixels.

How the Luminance and Chrominance Samples are Related in Space

As mentioned earlier in this chapter, the chrominance samples are sub-sampled by a factor of two in both the horizontal and vertical directions. This means that the spacing between chrominance samples is twice that of the luminance samples. So where are the chrominance samples located with respect to the luminance samples?

Take a look at *Figure 1-2*. For every four luminance samples, there's one Cb sample and one Cr sample. The Cb and Cr samples share the same location. The location of the chrominance samples is displaced vertically from the original sample point, so that it lies halfway between samples on two lines. Because of this scanning method, the chrominance samples have to be interpolated.

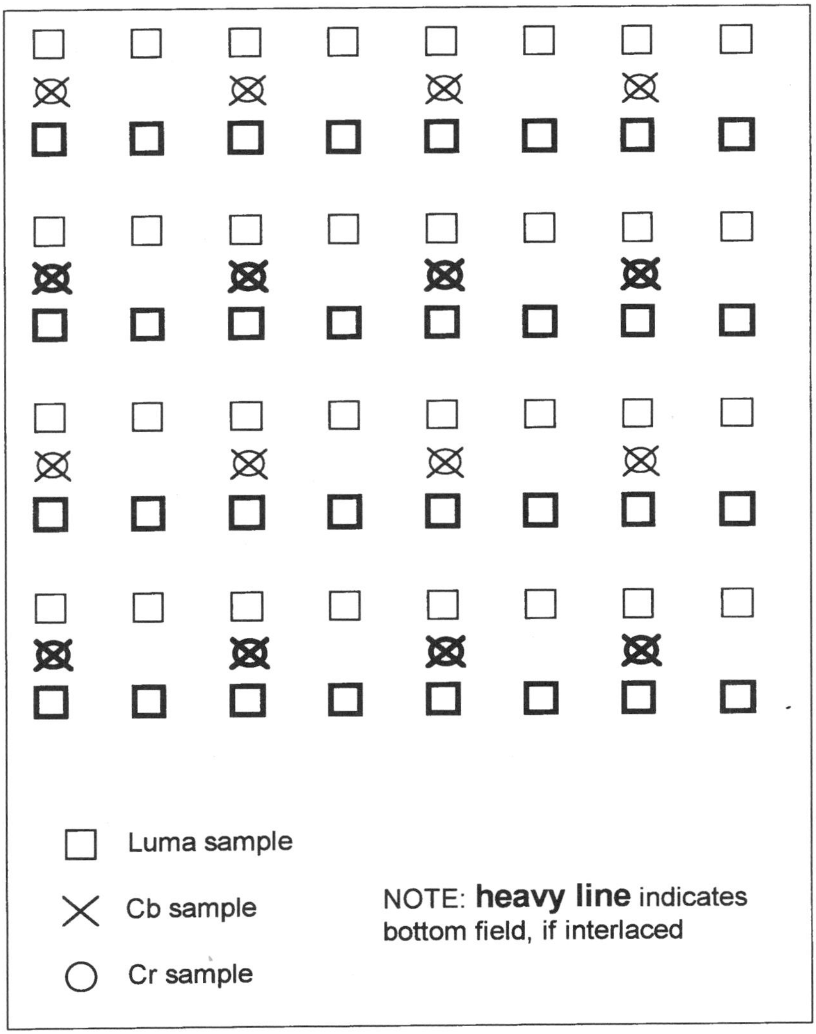

Figure 1-2

If the picture is one that is progressively scanned, the approach to perform this interpolation might simply be to average the two adjacent values, one above the luminance sample and one below the luminance sample, to compute the sub-sampled values. When the picture is to be interlace scanned, the vertical positions of the chrominance sample are not halfway between the luminance samples of the same field.

The reason that the process is performed in this way is so that the locations of the chrominance samples in space, within the frame, are the same for both interlaced and progressive sources.

Pixel Blocks

To keep things organized as the pixels are sent for further processing, groups of pixels are arranged into *blocks*. A *block* is an array of pixels. Alternatively, depending on the processing method, a block may be an array that consists of pixel values that have been *transformed* in some way. An *array* is a grouping in space that consists of rows and columns of values. In the case of the advanced television system, a block is an array of 8 lines by 8 pixels that represents either luminance or chrominance information.

Macroblocks

To further organize processing of pixel information, these 8 by 8 blocks are arranged into *macroblocks*. A macroblock is made up of four blocks of luminance values and two chroma (Cb and Cr blocks). A macroblock is therefore equivalent to a portion of the picture that is contained in an area of 16 pixels by 16 lines. The term *macroblock* may refer directly to pixel information, or it may refer to the transformed and coded representation of pixel data. *Figure 1-3* shows the composition of blocks and macroblocks. Since each block, either luminance or chrominance, contains an array of 64 (8 x 8) samples, and a macroblock consists of four luminance blocks and two chrominance blocks, a macroblock contains (4 x 64) + (2 x 64) or 256 + 128, for a total of 384 samples.

Lets take a look at how the macroblocks are arranged, then, in each of the ATV systems.

The 1080-line Format

Picture the 1080-line advanced television format like this: there are 1080 lines of pixels that make up the picture on the screen. Each line consists of 1920 samples. Actually, for coding purposes there are actually 1088 lines. The additional 8 lines are dummy lines added to make 1088.

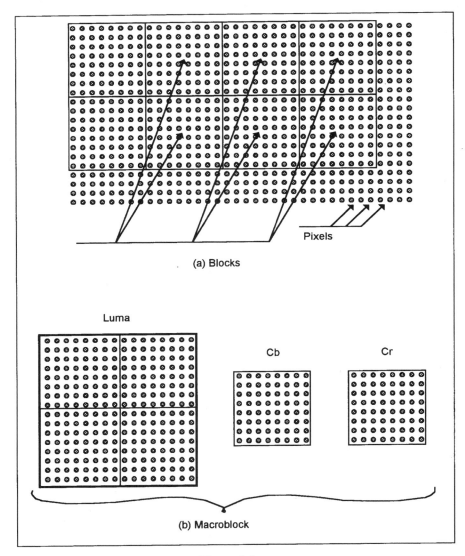

Figure 1-3

Remember, each macroblock is 16 pixels vertically, by 16 pixels horizontally. If you divide 16 pixels into 1088 lines, you will come up with 68. This is the number of macroblocks that make up the vertical dimension of the screen. Dividing 1920 samples per line by 16 gives the result of 120. Thus in the 1080-line format the image consists of 68 rows of macroblocks with 120 macroblocks in each row.

The 720-line Format

Let's perform the same math on the 720-line format. In this format there are 720 lines with 1280 samples in each line. Because, unlike the 1080 format, 720 is evenly divisible by 16, it is not necessary to add dummy lines. Thus 720 lines divided by 16 pixels vertically per macroblock yields 45. In the horizontal direction, 1280 samples divided by 16 pixels horizontally per macroblock yields 80. Thus in the 720-line format, the image consists of 45 rows of macroblocks by 80 macroblocks per row.

The 480-line Format

The image in the 480-line ATV image consists of 480 lines vertically, with 640 samples in each line. Performing the math in the same way as we did for the other two formats reveals that an image in this format consists of 30 rows of macroblocks with 40 macroblocks in each row.

Slices

Once the image that is to be transmitted has been neatly arranged into horizontal and vertical arrangements of macroblocks, it has to be encoded into a series of digital bits (called a *bit stream*) that are then processed by the remainder of the transmission end circuitry. Then it is sent out through space where it is received by a viewer's receiving TV set. The bits that the receiving set have to be decoded and reassembled into something that is identical, or nearly so, to the image that was sent.

Because of the nature of transmission systems, errors can be introduced into the digital bit stream. These errors, if allowed to propagate, could result in a garbled picture, or more likely because of the nature of the ATV system, loss of picture for periods that might be as brief as one field or frame. So, the system is designed so these errors will not be allowed to propagate very far.

The use of *slices* provides the ATV system with a mechanism for limiting the propagation of errors. Slices are created by grouping two or more adjacent macroblocks within the same row. The order of the macroblocks in a slice is from right to left, the same direction as the conventional raster scan.

To understand what this is all about, we have to consider the idea of a start code. A *start code* is a unique 32-bit code embedded in the coded bit stream. Start codes are used for several purposes, including telling the decoder what it's supposed to do with the series of bits that follows, or according to the glossary of terms "identifying some of the layers in the coding syntax." A start code consists of a 24-bit prefix (0x000001) and an 8-bit stream_id.

Before going any further, here's another concept: the coded bit stream that represents the images and sound that were in the original program material that's sent through space, consists mostly of digital codewords of variable length. If some kind of error is introduced between the coder at the transmitting end and the decoder at the receiving end, the output of the decoder might not be an accurate representation of the original program material. And this error might continue to cause inaccuracies in the program material until an event in the bit stream comes along to correct the problem; a new start code to put the coder back on track.

The slice is the minimum unit for resynchronization after an error.

The use of slices combats this propagation of errors by giving the decoder that information that's required to get things back on track. In essence, when an error does occur in the data stream, the decoder can wait for the start of the next slice, where the start code at that point in time tells it how to resume correct decoding.

The number of slices that the decoder creates represents something of a trade-off in terms of the efficiency of the compression of the material. The more slices that are used to make up the data stream, the better will be the error recovery. But at the same time, the larger the number of slices, the more bits get used that could otherwise be used to improve picture quality.

In the advanced television system, the first macroblock of each row of macroblocks is the beginning of a slice. There may be a number of slices in any row of macroblocks.

MAKING MOTION FROM SEQUENCES OF STILL PICTURES

The basic coding unit of a video sequence is the individual *video frame*. This is essentially a still picture. This picture is made up of the collection of slices that create the active picture area.

A *video sequence*, in other words a moving picture, is made up of two or more consecutive frames. In the coded bit stream, a video sequence contains a code at the beginning that tells the decoder that a succession of images that makes up a video sequence follows. This code is called a *sequence header*. At the end of the sequence, an *end-of-sequence* code in the bit stream tells the decoder that the sequence is complete.

Any video sequence may contain additional sequence headers within the sequence itself. Any of these sequence headers gives the decoder the information it needs to correctly determine what to do with the data in the bit stream that follows. Any point in the data bit stream that provides the decoder with information that enables it to become properly initialized and to deal with the information that follows is called an *entry* point. Decoding the stream of data that comes into the decoder and processing it into the video and audio information needed by the remainder of the advanced television circuitry is called *parsing the bit stream*.

VIDEO FRAMES AND THE TYPES OF VIDEO FRAMES

Figure 1-4 is a representation of a sequence of video frames that makes up a portion of a video program. Notice that they're not all labeled the same way. Some are labeled as I-frames, some are labeled as B-frames, and some are labeled as P-frames. It is the creation of these different types of frames that accounts for part of the compression in the video compression process.

The I-frames are referred to as *intraframe* coded frames. The P-frames are referred to as *predicted* frames. The B-frames are referred to as *bidirectionally* predicted frames. Let's take a look at what each of these types of frames is all about.

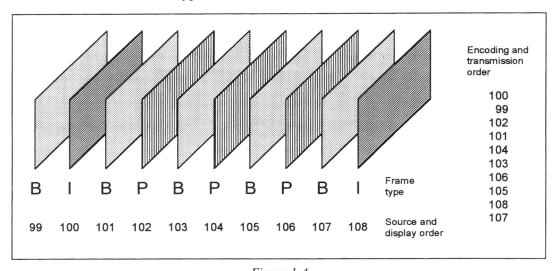

Figure 1-4

I-Frames

When we talked about video compression at the beginning of this chapter, we mentioned that a portion of the compression process was realized by transmitting a signal that contained information that told the decoder at the receiving end only what had changed in the scene since the last frame was transmitted, not the entire scene. This saves quite a bit of information channel bandwidth.

However, it also creates something of a problem. If there has been any kind of error in the transmission channel, the receiver might not be able to relate the frame change information to a previous frame. More important, if the set has just been turned on, or if the viewer has just changed the channel, there are no previous frames available. In any of these cases, the decoder would have no type of reference from which to begin proper decoding again, and the receiver's picture would probably fail completely.

To provide the receiver with some baseline type of information from time to time to allow it to recover from an error, some of the frames that are sent in a video sequence must be able to stand on their own, and not be dependent on previous or future frames. These frames are referred to as *intraframe* coded frames (intra- being the prefix meaning within), or *I-frames*.

The inclusion of I-frames from time to time in the transmitted bit stream gives the decoder in the receiving television set the information it needs to recover from an error, or to initialize and acquire the channel when the set is first turned on, or the viewer changes the channel. To sum it up, because the HDTV system is based on what is called "motion compensated prediction," an intraframe coded frame must be presented to the decoder to start the prediction loop. Therefore, I-frames are built into the system so that if the decoder loses synchronization for any reason, it can reacquire tracking quickly.

The intervals at which I-frames appear in the signal may vary. This parameter is selected at the encoder. This ability to vary these intervals allows consideration to be given to the need for random access, and the location of abrupt changes of scene in the video sequence.

P-Frames

In just a little moment, we'll get into where this is all going, and, it is hoped, this will all become clear, but for now we have to consider a couple of definitions, the meaning of which may not be clear yet. The definition of P-frame in this section, and B-frame, which follows immediately, are two of those terms.

P-frames are *predicted* frames. Moreover, they are frames that are predicted in the forward direction only. In other words, they are frames, images, that the system has generated based on the frame that came immediately before it. It is these predicted frames that allow the use of *inter*frame (*between* frames) coding techniques to improve the compression efficiency and picture quality. P-frames may include portions that are only intraframe coded. Each macroblock within a P-frame can either be intraframe encoded, or forward predicted.

B-Frames

Here's a term that's almost self-contradictory, even nonsensical, until you have a little more information. A *B-frame* (the B means that it is *bidirectionally* predicted) is a type of frame that is predicted based not only on frames that have already occurred, but are also predicted based on future frames: frames that, apparently, have not occurred yet. The future or previous frames on which a B-frame is predicted, sometimes called "anchor" frames, will always be either an I-frame or a P-frame.

The basis of the B-frame prediction is that a video frame is related to the frames that comes before it, and the frames that comes after it in the video sequence. Therefore, if a future frame is available to the decoder in the receiving set (as well as the previous frame), the decoder can do a better job of predicting a frame than it could if only the previous frame were available.

The preceding discussion certainly brings up an interesting question: "how does the receiving set decoder gain access to a frame that hasn't been sent yet." The answer is: "it doesn't." The secret is a certain type of manipulation of the image information that is not possible in the NTSC system: the transmission system sends the images in a different order than the order in which the receiving set will display them.

Obviously then, the encoder at the transmitting end spends a little extra time waiting for more than one frame so it can change their order when it is useful to do so, and the decoder at the receiving end, has to spend a little extra time waiting for more than one frame so it can perform a more accurate prediction of a B-frame, and then put the frames back in their original order.

To restate everything as the Advanced Television System Committee document puts it: "... consequences of using future in the prediction are: the B-frame cannot be used for predicting future frames, the transmission order of frames is different from the displayed order of frames, and the encoder and decoder must reorder the video frames, thus increasing the total latency."

In *Figure 1-4*, there is one B-frame between each pair of I/P frames. Each pair of frames is labeled as to both its transmission order and its display order.

The reason that the ATV system uses B-frames is to improve the efficiency of the compression, and the picture quality perceived by the viewer when the delay, or latency, caused by this process is not an important consideration. The use of B-frames improves coding efficiency whether it is used in the case of interlaced or progressive scan. The number of B-frames that is transmitted between any two reference frames (I or P) can be determined at the encoder.

MOTION ESTIMATION

This entire business of prediction, and all those types of frames can get a little confusing, so let's pause here to consider a concrete example with which to consider the idea of motion estimation. A relatively simple example. Let's say that we're working with a motion picture that consists of a truck driving down a road. We're looking at it from a motionless camera, and the truck is moving from left to right. To further simplify things, let's say that the scene is in the desert on a very clear day with no clouds or motionless clouds. So the only thing that's moving is the truck. And it's not even kicking up any dust.

Let's say that this sequence has been going on for a short while, and start when the encoder generates an I-frame, and let's not consider B-frames at the outset. The encoder generates an I-frame and transmits it via the airwaves, and the signal is received by a receiving HDTV set. The image is displayed, but the decoder retains the information contained in that image.

Now the encoder generates a P-frame based on the motion of the truck. It generates a frame that shows the background that looks just like the background in the preceding I-frame. The truck has moved a little to the right. Now the encoder compares the image it has predicted with the actual image that follows. This is based on some abstruse math that we won't go into here.

The result of this comparison is data that represents how far from the actual image the predicted image is.

Now keep in mind that the decoder at the receiving set performs its magic using the same math that the encoder uses. It is also generating a prediction of what the next frame will look like. The only information it needs to make the next P-frame very nearly like the image that is at the transmitting end is the data that tells it what the error in prediction at the transmitting end was. In the case of the truck moving at constant speed against a still background, that data will take up very little bandwidth. If the transmission was a still picture, the data that the receiver needed to generate P-frames would be minimal.

Let's say now, that the scene shifts to show the driver behind the wheel of the truck. In this case, the image has changed completely. The value of every, or almost every, pixel in the image has changed. This more or less completely throws off the entire prediction process, and now the difference between the predicted image and the actual image is huge, and all of that difference information has to be sent, at the expense of reducing the compression, for that period of time.

The same type of situation exists if the motion in a scene were extreme, as when a camera is panning to follow a speeding car, or if it is necessary to transmit sharp detail in a scene with motion.

The Vector Search Algorithm

The ATV video coding system uses *motion compensated prediction* as part of the data compression process. In this motion compensated prediction, macroblocks in the frame that is being generated are predicted by macroblock-sized regions in frames that have been transmitted previously. The term "motion compensation" describes the fact that the locations of the macroblock-sized regions in the frame that is being used as a reference can be offset to create the effect of motion in that area of the frame. As when the truck in the earlier discussion moved from region to region of the images in the succession of images that made up the video sequence. The direction and distance of movement of the microblocks are known as *motion vectors*.

The Prediction Loop

We've discussed the concept of prediction of the next frame in a succession of video frames based on the previous frames. This prediction is performed in the encoder prediction loop, *Figure 1-5*. This closed feedback loop is the entity within which the bulk of the compression of an ATV program takes place.

This loop includes a prediction function that predicts, or estimates, the picture values of the next image to be encoded in the sequence of successive pictures that is available within the loop, based on previous images. Keep in mind that an identical loop exists at the receiving end. Thus, if the prediction loop at the transmitting end successfully predicts what the next image will look like, we can conclude that because it had essentially the same information, the prediction loop at the receiving end has been similarly successful. (Errors caused by problems in the transmission medium will cause anomalies at the receiving end, but, based on tests and experiments, such errors are expected to be rare.)

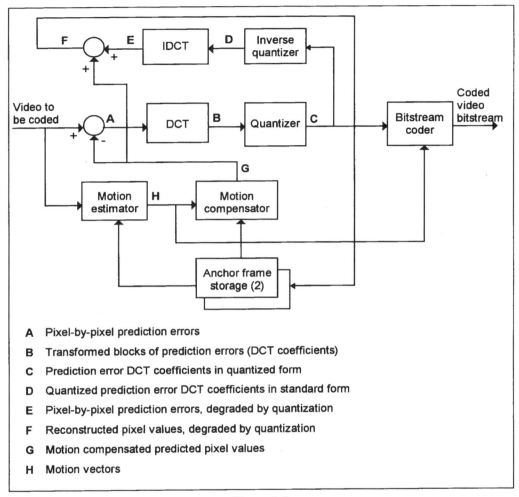

A Pixel-by-pixel prediction errors

B Transformed blocks of prediction errors (DCT coefficients)

C Prediction error DCT coefficients in quantized form

D Quantized prediction error DCT coefficients in standard form

E Pixel-by-pixel prediction errors, degraded by quantization

F Reconstructed pixel values, degraded by quantization

G Motion compensated predicted pixel values

H Motion vectors

Figure 1-5

Once the prediction loop has generated the predicted image, the picture values in this image will be subtracted from the picture values in the actual subsequent picture. This part of the process is considered to be at the core of predictive coding. The goal is to predict the subsequent image so well that the result of the subtraction of it's picture values from those of the actual subsequent image is zero, or close to zero, in most cases.

These prediction differences are calculated separately for the luminance and the two chrominance components before further processing.

When it is deemed necessary to send an intraframe encoded frame (I-frame), prediction is not used for an entire frame, or part of a frame.

Spatial Transform Block

The difference between the values in the predicted frame and the actual frame (these are sometimes referred to as prediction errors) are grouped into 8-by-8 blocks, and a spatial transform is applied to the blocks of difference values. In the case of an intraframe-coded frame, the spatial transform is applied to the data from the new frame. The luminance and the two chrominance values are transformed separately. Because the chrominance data is sub-sampled vertically and horizontally, each 8-by-8 block of chrominance data (Cb or Cr) corresponds to a 16-by-16 block of luminance data.

The spatial transform used is called a *discrete cosine transform* (DCT). This is a very involved mathematical procedure, which is well beyond the scope of this treatment of ATV.

Quantization

Transformation of the original data using the DCT causes the data to be organized in such a way that emphasizes the spatial frequency components of the images or image differences. In other words, it highlights to the encoder if an area of the image is relatively "smooth" visually, or if there are a lot of differences or appearance in the area: the area is "busy." The encoder has been provided data that describes the response of the human visual system to various spatial frequencies. This allows the encoder to use adequate precision to encode the image so that the image recreated at the receiving end appears accurate to the human eye-brain system, but not to use any greater precision than necessary.

Let's say, for example, that an area of the picture is of a quilt with a tight, wild pattern, hanging on a clothesline, and beyond the quilt, the picture is a clear blue sky. Any inaccuracies in the transmitted data representing the sky will be obvious and make the picture seem odd. On the other hand, the strong pattern in the quilt will mask many deficiencies in the picture values.

Inverse Quantization and Inverse Spatial Transform (IDCT)

At the receiving set, the mathematical representation of the image, or image differences, are decoded and the inverse of the quantization process that took place at the transmitting end is imposed. These data are organized as 8-by-8 blocks of DCT coefficients (a portion of the mathematical representation of the image) and the inverse discrete cosine transform (IDCT) is applied to each block. This results in a new array of pixel values, or pixel differences, that correspond to the output of the subtraction at the beginning of the prediction loop. If the prediction loop was in the interframe mode, the values will be pixel differences. If the loop was in the intraframe mode, the inverse transform will produce pixel values of an entire frame.

IMAGE REFRESH

The foregoing text described how an image may be sent by transmitting the information that tells the decoder in the receiving set how the new image differs from the previous image sent. This works fine, except when the set is first turned on, or the channel has been changed, or there has been a transmission error, and there is no valid initial image to update. Moreover, it is not desirable to have too many predicted pictures in succession, to control the adding up of errors caused by *IDCT mismatch*.

The video coding system does not completely specify the results of the IDCT operation. This is intentional. Because of this characteristic, however, successive images may become further and further different from the actual image, even if the transmission system hasn't introduced any errors. The system keeps this "drift" from becoming too extreme by requiring that each macroblock be coded without prediction (intracoded) at least once in any 132 consecutive frames.

In order for a decoder to become initialized, or re-initialized, it must have received valid picture data that is not described with reference to picture information that was previously transmitted. This process is called *image refresh*. In other words, image refresh is accomplished using intraframe coded macroblocks. Image refresh can be accomplished in one of two ways: periodic transmission of I-frames, or progressive refresh. A combination of these two types of image refresh may also be used.

Periodic Transmission of I-Frames

One way to accomplish image refresh is to transmit an I-frame from time to time; that is, a frame that has been entirely intraframe encoded. These intraframe coded frames may be sent at regular intervals, or the frequency of their inclusion may be varied.

When this scheme of image refresh is used, the receiver may go through this sequence:

• Ignore all data until receipt of the first sequence header.

• Decode the sequence header and configure circuits based on sequence parameters.

- Ignore all data until the next received I-frame.

- Commence picture decoding and presentation.

It is possible that an I- or P-frame will contain uncorrectable errors, possibly introduced by a fault in the transmission. When this occurs, there will probably be a picture error that propagates, because subsequent frames are predicted on the basis of an image that contained an error. This error will probably continue to be propagated until the receiver receives an error-free *I-frame*.

Progressive Refresh

There is another way to achieve an image refresh, called progressive refresh. In this process, portions of successive frames are intraframe encoded. When this method of image refresh is used, it is imperative that the areas of the image are chosen so that within some reasonable number of frames, every macroblock of the frame is intraframe encoded at least once.

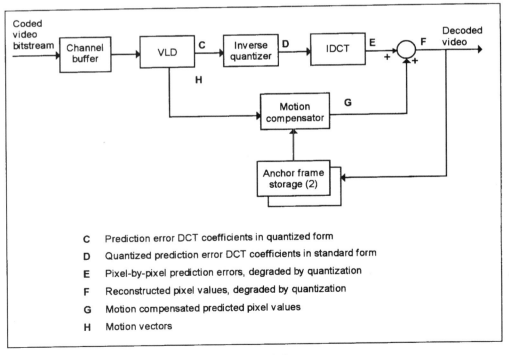

C	Prediction error DCT coefficients in quantized form
D	Quantized prediction error DCT coefficients in standard form
E	Pixel-by-pixel prediction errors, degraded by quantization
F	Reconstructed pixel values, degraded by quantization
G	Motion compensated predicted pixel values
H	Motion vectors

Figure 1-6

THE DECODER

Most of what we have been talking about so far has described the transmitting end. That's because it's important to know what type of signal is being transmitted in order to understand the nature of the receiving equipment.

Figure 1-6 shows the video decoder. Simply put, the video decoder consists of circuitry that performs the inverse of the processes that were performed in the encoder. Once it has been selected by the tuner, the received encoded video bit stream is placed in the channel buffer. Bits are taken from the buffer, and started through the process of reconstruction of the image information: either pixel values, or prediction errors.

When decoding an interframe coded frame, the decoder uses the motion vectors pertaining to that frame with reference to the previous frame to perform a prediction just exactly as the encoder did it. The prediction errors are summed with the results of motion compensated prediction to produce pixel values. The decoder is provided with the capability to minimize the perceived degradation when transmission errors occur.

Frame Store for Decoded Pictures

When an anchor frame (I-frame, or a P-frame) has been decoded from the incoming bit stream, it is going to be used as a reference for future predictions. Therefore, in addition to being displayed as images on the screen, they must be temporarily stored in a *frame buffer* until they have been used for prediction. B-frames must also be stored temporarily so that the correct order of frames that have been received out of order may be restored.

2
AUDIO IN THE ATV SYSTEM

HDTV is designed to provide viewers with the whole home theater experience. That means that in addition to the high definition video picture, there will be six channels of high fidelity audio for stereo surround sound. This is frequently referred to as 5.1 channels, because the sixth channel is generally limited to the frequency range from about 3 Hz to 120 Hz for driving the subwoofer (also called "low-frequency enhancement, or LFE for short). This system as it appears at the receiving end will be discussed in more detail.

The audio subsystem in the ATV system consists of the complete audio path, from the point where it enters the audio encoder(s) at the transmitting end, to the point where is emerges from the audio decoder(s) at the receiving end. The audio subsystem is shown in *Figure 2-1*.

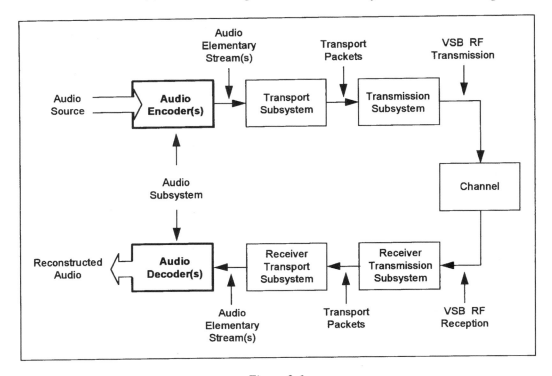

Figure 2-1

AN OVERVIEW OF THE ATV AUDIO SYSTEM

As with the video in the ATV system, the audio is also converted from the original analog audio signal into a digital signal to be further processed and transmitted. The process of converting the audio from analog to digital is called "encoding." As shown in *Figure 2-1*, the audio subsystem consists of elements at the transmitting end, and at the receiving end. At the transmitting end, the audio subsystem element comprises the encoding function, and is between the audio inputs and the transport subsystem. At the receiving end, the audio subsystem element comprises the audio decoding function and is placed between the receiver transport subsystem and the audio output.

The audio encoder(s) generates the "audio elementary stream(s)." The audio elementary stream is a string of digital bits that is a digital representation of the baseband audio input signal. So as not to introduce unnecessary confusion at this point, let's define "baseband." It simply means that the audio is a standard audio signal, 20 Hz to 20 kHz, or thereabouts, the way is came from the microphone or some audio source, that has not in any way been modulated or modified.

The transport system is designed to be flexible so that it can handle a single audio elementary stream, or a number of audio elementary streams. The transport subsystem at the receiver is responsible for selecting which audio stream or streams to deliver to the audio subsystem. The audio subsystem at the receiver decodes the audio elementary stream or streams back into baseband audio for delivery to the audio system.

If the television station merely transmitted this audio elementary stream representing the audio portion of a TV program, the receiving set would see that here was a string of digits, but wouldn't know what it was for. In order for the receiver to process this data as audio, the stream has to be identified as such. For that reason, the audio elementary stream is grouped into "packets," by the transport subsystem. These are known as PES (packetized elementary stream) packets. As will be described in more detail later, these packets contain not only the stream of digits, but other information as well, that tells the receiving end what kind of information this is, and what to do with it.

The PES packets are further packetized into transport packets. At the transmission subsystem, the transport packets are converted into a modulated RF signal for transmission to the receiver. When the signal reaches the receiver, it is demodulated by the receiver transmission subsystem. The receiver subsystem converts the received audio packets back into an audio elementary stream, which is decoded by the HDTV audio decoder.

The block diagram of the audio subsystem in Figure A1 does not necessarily represent the way in which an actual audio subsystem will function. It is merely intended to express the concept. Actual audio subsystems may differ from this idea. For example, the transport processing may be divided into two blocks: One to do the PES packetization, and another to do the transport packetization. Or some of the functions of the transport subsystem may be included in either the audio coder or the transmission subsystem.

The Audio Encoder Interface

That audio elementary stream we were talking about can actually contain six audio channels of audio programming. As we mentioned before, those six channels (or 5.1 channels if you prefer) are: Left, Center, Right, Left Surround, Right Surround, and Low Frequency Enhancement (LFE). Moreover, the transport can actually transmit more than one of these audio elementary bit streams.

The bandwidth of the LFE channel is generally limited to the range of 3 kHz to 120 kHz.

Audio Sampling

The system transmits digital audio sampled at a frequency of 48 kHz, locked to the 27 MHz system clock. The sampling rate at the input to the audio encoder must be locked to the video clock in order for the audio subsystem to operate properly.

Audio Compression

When we talked about the video portion of the ATV signal, we talked about the importance of compressing the video so that it could be transmitted in a 6 MHz television channel. Compression of the audio portion of the program is also desired, for two reasons. One reason for audio compression is so that the channel bandwidth can be used most efficiently. The second reason is that it reduces the amount of medium required to store the program material. The aim of audio compression is to be able to represent an audio signal with as few bits as possible while maintaining the level of sound quality consistent with the particular application.

The digital audio compression standard used in the ATV system is specified by the Advanced Television System Committee Document identified as A/52, and is referred to as AC-3.

For the purposes of the ensuing discussion, let's broadly define a term here: "psychoacoustics." At the risk of oversimplification, the term psychoacoustics suggests the mind's interpretation of the audio information presented to the ears. On example of psychoacoustics is what a listener experiences when he sits midway between the two front speakers and listens to a stereo program. It might sound as though the players in the orchestra playing the music being reproduced from, say a CD, are arrayed in a stereo "sound stage." It sounds as though the violins are over there on the left, the bass viols are over there on the right, and the voice of the singer is right there in the middle in front of the listener.

In a more spectacular scene from a movie in a home theater, the listener might hear the sound of a train coming in from the left as the video shows it arriving, and he would hear the sound move from left to right following the motion of the train on the screen.

As another example of psychoacoustics, if someone coughs when everything else is silent, the cough might sound quite loud and obtrusive. If someone coughs while that train is going by, no one would even hear it. The process of audio compression takes psychoacoustics into account.

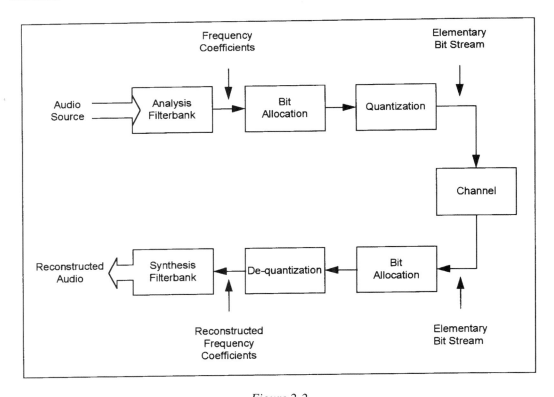

Figure 2-2

The audio compression system consists of three basic operations (*Figure 2-2*). In the first stage, the representation of the audio signal is changed from the time domain to the frequency domain (see the accompanying explanation of time domain versus frequency domain). It is more efficient to perform psychoacoustically based audio compression in the frequency domain than in the time domain. The result of this conversion is a mathematical expression describing the frequency domain representation of the digitized audio signal. The coefficients of each of these expressions in time are what are then encoded.

When an analog parameter is converted into its digital equivalent, the process of that conversion is called "quantization." If it's important to accurately represent any given value of the analog parameter, then it will require a great number of digital bits to do so. For example, a binary number of four bits can represent any decimal number from zero to 1 (00000 to 1111), whereas eight binary digits can represent any binary number from 0 to 255 (00000001 to 11111111). If you're going to use four bits to quantize an analog value, one bit can represent 1/16 of the maximum value, whereas using eight bits to quantize an analog value one bit will represent 1/256 of the maximum value.

In sum, the more bits used to represent a changing quantity, such as an analog signal, that has been digitized, the smoother, and more accurate will be the signal when it is converted back to its analog form. However, the more bits used to represent that quantity, the more bandwidth required to store and transmit it.

If you use fewer bits to represent the signal, you can use less bandwidth to store and transmit it, but you increase "quantization noise," the error in the reproduced signal, which is perceived as noise. In the case of audio as handled in the ATV system, the frequency domain coefficients can be quantized with relatively few bits (coarsely quantized), because the quantizing noise will be at the same frequency as the audio signal, and therefore relatively low signal to noise ratios (SNR) are acceptable because of the phenomenon of "psychoacoustic masking:" the desired audio signal covers up the undesirable noise.

A psychoacoustic model of human hearing has been set up, and is used in the audio subsystem of ATV. The number of bits that will be used to represent a particular portion of the audio is determined by the audio subsystem based on the SNR acceptable in accordance with this model. In other words, the system determines, on the fly, just how many (or more accurately, how few) bits are required to digitize a portion of the audio program without causing degradation of the audio output at the receiver, and uses just that many bits in encoding. In the final step, the frequency coefficients are quantized to the precision required, and formatted into the audio elementary stream.

The Audio Sync Frame

The fundamental unit of encoded audio is the *AC-3 sync frame*. Each sync frame of audio is a completely independent coded entity that consists of 1536 audio samples. The elementary bit stream contains the information necessary to allow the audio decoder to determine how many bits were used by the encoder to quantize each sample, which allows the decoder to unpack and convert the quantized elementary bit stream coefficients back to their analog equivalents. The synthesis filterbank is the inverse of the analysis filterbank. It converts the reconstructed frequency coefficients back into a time domain signal.

Bit Stream Syntax

The system at the transmitting end performs a lot more mathematical processes on the audio before it reaches the airwaves, to construct the audio signal and further compress it, but it's not really necessary to bother with that to get an idea of what the transmitting end audio sends out. So we'll go right into a look at the audio bit stream.

The audio bit stream consists of a repetition of audio frames which are referred to as AC-3 sync frames (*Figure 2-3*). Each AC-3 sync frame is self-contained. It consists of synchronization information, (SI), bit stream information (BSI), 32ms of encoded audio, and a CRC (cyclic redundancy check) error check code. That's an interesting number. It takes about 1,000 audio frames to generate 32 seconds, or about a half-minute of audio.

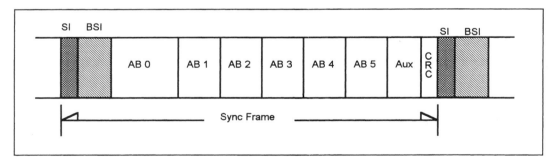

Figure 2-3

Every one of these sync frames contains the same number of bits, and contains six encoded audio blocks. Just as the digitally encoded video frame was called the video "access unit," the sync frame may be considered the audio access unit. The SI, that is system information, portion of an audio frame contains the following information:

- a 16-bit synchronization word.

- an indication of audio sample rate 48 kHz for the digital television system).

- an indication of the size of the audio frame (which indicates bit rate).

Loudness and Dynamic Range

Most viewers are annoyed when they've turned up the volume on their sets so they can hear a quiet program, only to be jangled when a loud, blaring, commercial comes on, or when they change channels and the new program is much louder than the one they had been watching. The ATV system makes it possible to minimize the occurrence of this type of situation.

Most audio programs involve the human voice. One way to make various programs have similar loudness levels is to make the loudness of dialogue (spoken in a normal voice; not shouting or whispering) approximately equal on all channels and in all programs. The AC-3 audio system provides methods of making this an achievable goal.

This is a good place to talk about the audio parameter called "dynamic range." Dynamic range was really not much of an issue until the introduction of compact discs (CDs), and then other forms of digital music. Dynamic range is simply a measure of the difference between the softest and the loudest passages in an audio program. The sound levels on a vinyl LP disc (for those of you who remember them), or in the NTSC TV system, were such that if you turned up the volume control so that you just barely heard the softest passages, the loudest passages might sound loud, depending on the program material, but loud within reason.

The digital audio coding system, on the other hand, is capable of delivering more than 100 dB of dynamic range. If the softest sounds in a program were barely a whisper, and the loudest passage was 100 dB above that, the loudest sounds would be, literally, deafening.

At the moment, there are no regulations that limit how loud dialogue may be in an encoded audio bit stream. However, because of the broad dynamic range available with digital audio, there is no technical reason for dialogue to be encoded near 100 percent, as is commonly done in NTSC television.

However, there is no assurance that all channels, or all programs or program segments on a given channel will have dialogue encoded at anywhere near similar levels. Since there is no uniform coding level for dialogue, there would be fluctuations in audio level between programs.

The AC-3 audio standard includes a mechanism by which the sound level of various TV programs can be equalized. Encoded elementary bit streams in the AC-3 system include a tag, that indicates the subjective level at which dialogue has been recorded. This parameter is called dialnorm. Different audio programs may be encoded with different amounts of dynamic range ("headroom") above the level of dialogue in order to allow for dynamic music and sound effects. DTV receivers are able to use the value of dialnorm to adjust the reproduced level of audio programs so that different programs received by the set will be able to reproduce spoken dialogue at a uniform level.

Depending on how the engineers designed them, some receivers may offer the listener a volume control that is calibrated in absolute sound pressure level. With an audio system like this, the viewer could dial up the desired sound pressure level for dialogue, and the receiver would adjust the level of every decoded audio program so that the dialogue is always reproduced at the desired level.

The portion of the sync frame that contains the bit stream information contains the dialnorm field, a five-bit field that indicates the average level of dialogue within the encoded audio program. This value is relative to the level of a full-scale 1 kHz sine wave. The measurement of dialogue sound level is performed using a method that gives a value that accurately represents the sound level as perceived by the listener: it is accurate *subjectively*.

As is implied by the term *subjective*, this measurement is not an exact science, and it is anticipated that new measurement techniques for this audio information will be developed in the future. The measurement currently used, and which will be continue to be used until a better measure is developed is the "A" weighted integrated measurement (L_{aeq}).

If broadcasters and other entities that deliver encoded audio bit streams don't ensure that the value of this important parameter, dialnorm, is correct, there will be some unusual sound transitions in the homes of some viewers. Some unscrupulous TV executives, or perhaps careless, or clueless engineers, could set up a commercial message so that dialnorm would be set to indicate dialogue at a low level, but the dialogue was actually set at full level. In this case, the commercial would be reproduced at the same level as an explosion in a feature film: greater than 100 dB sound pressure level in some home theater setups. That would be startling, to say the least, to the viewers. There are methods by which broadcasters can ensure that this does not happen.

Dynamic Range Compression

Most high-quality programming is produced so that the audio portion has a broad dynamic range that is suitable for the highest quality audio reproduction. Because broadcasters service a wide audience—viewers with small monophonic speakers as well as viewers who have home theater equipment, and everything in between—they typically limit the dynamic range of the audio they broadcast. In the case of NTSC, all viewers receive the same audio, which has limited dynamic range, so that it is not possible for viewers with the latest audio equipment to enjoy the original audio with its broad dynamic range.

It's a different story with digital TV. The audio coding system in the advanced TV system provides a dynamic range control system embedded in the audio signal that allows a common encoded bit stream to deliver programming with a dynamic range appropriate for each individual listener. Within each audio block, every 5 ms, the HDTV broadcast signal provides a dynamic range control value (dynrng). These values are used by the audio decoder at the receiving set to adjust the level of the reproduced audio for each audio block. Level variations of up to +/- 24 dB may be indicated.

The values of dynrng generated are such that the broadcast audio signal will provide an audio program that is subjectively pleasing, but restricted in dynamic range. If the receiving set is not provided with some means of adjusting the dynamic range, the dynamic range of the audio will be dialogue level.

For sounds quieter than dialogue values, values of dynrng will indicate an increase in gain. For sounds louder than dialogue values, values of dynrng will indicate a decrease in gain. In short, the broadcaster is in control of the values of dynrng, and can supply values of dynrng that the broadcaster thinks are appropriate. Using dialogue level as the level at which the gain is unaffected contributes to the uniformity of the program loudness.

If no action is taken by the viewer, or allowed by the receiver circuitry, the audio decoder will use the values of dynrng imbedded in the audio signal. The result will be that the receiver will produce an audio program that is essentially that dictated by the broadcaster—a program with limited dynamic range.

Some receivers may offer an option to the viewer to scale the value of dynrng in order to reduce the effect of the dynamic range compression that was imposed by the broadcaster. In the extreme case, if the viewer scales the value of dynrng to zero, the audio will be reproduced with the same full dynamic range as that of the original program material.

Optional scaling of dynrng can be done differently for values that indicate gain reduction (reducing the level of loud sounds) and for values that indicate gain increase (making quiet sound louder). A system such as this gives the viewer independent control of compression applied to loud sounds and quiet sounds. The result of this control is that although the broadcaster imposes dynamic range compression to suit the needs of most of the audience, individual listeners may have the option to enjoy the audio program in whatever dynamic range they wish.

Additional Audio Programs

One AC-3 audio elementary stream contains the encoded representation of a single audio service; involving, of course up to six (or 5.1) audio channels. It is possible for the broadcaster to deliver more than one audio service, by broadcasting multiple audio streams. When the broadcaster transmits more than one audio service, each elementary stream is conveyed by the transport multiplex with a unique PID (packet identifier).

Each elementary stream may be coded to be recognized as one of several different types of audio service. This identification is carried in the bsmod bit field. Their are eight different types of service: two types of *main service* and six types of *associated service*. Each associated service may be identified as being associated with one or both main audio services. This identification is carried in the AC-3 audio descriptor in the transport PSI data. Each AC-3 elementary stream may also be identified with a language code.

Table 2-1 describes the various audio service types, and also shows what the value of bsmod that identifies the service type will be. These will be described in detail in paragraphs to follow. Associated services may contain complete program mixes, or may contain only a single program element. Associated services that are complete mixes may be decoded and used as is. These associated services are identified by a bit in the AC-3 descriptor identified as the full_svc bit. Associated services that contain only a single program element are intended to be combined with the program elements from a main audio service.

The complete audio program that is presented to the listener via the speakers may consist of any of a number of program possibilities. For example, the audio program might consist of a main audio service. Another possibility is that audio program might be from an associated audio service that is a complete mix. Another possibility is that the audio might consist of a main audio service in combination with one associated audio service.

In order to correctly reproduce the audio program from some audio service combinations that may be broadcast by transmitting stations, it may be necessary to decode one main service and one associated service simultaneously. It is important to keep in mind that some receivers may not have this capability.

bsmod	Type of service
000 (0)	Main audio service: complete main (CM)
001 (1)	Main audio service: music and effects (ME)
010 (2)	Associated service: visually impaired (VI)
011 (3)	Associated service: hearing impaired (HI)
100 (4)	Associated service: dialogue (D)
101 (5)	Associated service: commentary (C)
110 (6)	Associated service: emergency (E)
111 (7)	Associated service: voice-over (VO)

Table 2-1 Service Types

Service Types

Following are descriptions of each of the service types:

- *Complete main audio service (CM)*: This is the normal mode of operation. All elements of an audio program are available in this mode. The audio program may include any number of channels from 1 to 5.1.

- *Main audio service; music and effects (ME)*: In this service type, all elements of an audio program are present, except for the dialogue. This audio program may include from 1 to 5.1 channels. If the original program included dialogue, this dialogue may be provided by a D-associated service. In this case, the dialogue service will have to be decoded simultaneously and added to complete the program.

- *Associated service: visually impaired (VI)*: The VI service is typically a single-channel service, the purpose of which is to convey a narrative that describes what is going on in the picture. This service is decoded along with the main audio service. In other cases, the VI service may be provided as a complete mix of all program elements. In those cases, this service may use any number of channels, up to 5.1.

- *Associated service: hearing impaired (HI)*: The HI service will ordinarily be a single-channel service, intended to convey dialogue that has been processed to make it easier to understanding by a viewer who is hearing impaired. This service is decoded along with the main audio service. In other cases, the HI service may be provided as a complete mix of all program elements. In those cases, this service may use any number of channels, up to 5.1.

- *Associated service: dialogue(D)*: The D service conveys dialogue. This portion of the audio program is to be mixed into a main audio service that does not contain dialogue.

- *Associated service: commentary (C)*: The commentary service will generally be a single channel of commentary that may be decoded along with the main audio service at the option of the viewer. This service is different from that provided by the D service. The material provided by the D service is essential to the program. The material conveyed by the commentary channel is an optional addition to the program content. In some cases, the C service may be provided as a complete mix of all program elements. In those cases, this service may use any number of channels, up to 5.1.

- *Associated service: emergency message (E)*: This audio service will be used for bringing important information of an emergency nature to the viewers. This is a single-channel service that will be given priority in reproduction. If the receiving set recognizes that the audio service has this identification, it will route it directly to the audio decoder. Any time the decoder receives a portion of the program of this type, it will mute the main service and decode and reproduce the information carried by the E channel.

- *Associated service—voice-over (VO)*: This is a single-channel service. When this service is received by the ATV receiver it will be decoded and added into the information already headed to the center channel.

Multilingual Services

An audio bit stream that carries dialogue can be in any language. If a broadcaster wishes to provide audio services in many languages, it may provide a number of main audio services, each in one of the desired languages. In terms of providing realistic audio, this is the preferred method of delivering multiple languages, because it allows placement of the dialog without any restrictions, and includes an reverberation related to the dialog. However, this method has a major technical disadvantage: each language provided requires all 5.1 audio channels, and each such channel requires 384 kbps.

One way to reduce this bit rate is to reduce the number of audio channels for some of the languages broadcast. This makes sense if there are relatively few viewers who speak a given language. As an example, alternate language channels might be provided in 2-channel stereo. This would require a bit rate of 128 kbps. Even more conserving of channel bandwidth would be to transmit a monophonic version of an alternate language, and use a bit rate of only 64 kbps to 96 kbps.

There is another way to offer services in more than one language: broadcast a main multi-channel audio service (ME) without dialogue. Then a number of single-channel dialogue associated services (D) can be provided. Each of these channels could be provided at a bit rate of approximately 64 kbps to 96 kbps. If the broadcaster transmitted multiple language dialogue in this manner, reproduction of the complete audio program at the receiver would require that the desired language D service be simultaneously decoded and mixed into the ME service.

The advantage of this approach is that it allows a large number of languages to be provided with the most efficient use of audio bandwidth. The disadvantage of using this method is that it doesn't allow proper realistic reproduction of the sound. The audio effect would be that the single channel of dialogue would be mixed into the center channel, and therefore the sound would be fixed in space. Let's say that someone in a scene was moving and speaking at the same time. That person's voice would not follow the person, but would remain stationary in the middle of the soundstage. Moreover, the reverberation due to the dialogue would remain in the center of the audio program, which would further detract from the audio realism.

In spite of its drawbacks, this type of program would be adequate for some types of programming, such as sports. And it would be attractive to use this method because of the saving in bit rate it affords. However, as mentioned before, some receivers may not be provided with the capability to simultaneously decode an ME and a D service.

If a broadcaster wishes to efficiently broadcast stereophonic service in multiple languages with audio that will faithfully reproduce the original program material, they can do so by transmitting a stereo ME main service accompanied by stereo D services. The D and desired language ME services will then be combined in the receiver into a complete stereo audio program. Using this method, the dialog may be panned to follow the action on the screen, and reverberation can be included in both channels. A high-quality stereo ME service can be sent at 192 kbps. The accompanying stereo D services, consisting of voice only, can be transmitted at lower bit rates, such as 128 or 96 kbps for each language.

Another efficiency of this scheme is that during any portion of the program that does not involve dialog, the broadcaster can suspend transmitting the D service channels, and use that bandwidth for other purposes.

Audio Bit Rates

Table 2-2 is a general guide to the audio bit rates that are expected to be used by broadcasters during broadcast of an ATV program. For main services, the use of the LFE (subwoofer) channel is optional and won't affect the indicated data rates.

Type of service	Number of channels	Typical bit rates
CM, ME, or associated audio service containing all necessary program elements	5	320-384 kbps
CM, ME, or associated audio service containing all necessary program elements	4	256-384 kbps
CM, ME, or associated audio service containing all necessary program elements	3	192-320 kbps
CM, ME, or associated audio service containing all necessary program elements	2	128-256 kbps
VI, narrative only	1	48-128 kbps
HI, narrative only	1	48-96 kbps
D	1	64-128 kbps
D	2	96-192 kbps
C, commentary only	1	32-128 kbps
E	1	32-128 kbps
VO	1	64-128 kbps

Table 2-2

The maximum bit rate that must be decoded by the receiver determines the size of the audio decoder input buffer, and of course, therefore, part of the cost of that device. The AC-3 digital audio standard supports bit rates ranging from a minimum of 32 kbps to a maximum of 640 kbps for each elementary bit stream. The bit rate that was selected for the digital television system was restricted so that the size of the input buffer in the audio decoder could be reduced, and thus its cost would also be within reason.

Receivers can be expected to support the decoding of a main audio service, or an associated service that is a complete service, at a bit rate up to and including 384 kbps. Transmissions may contain main audio services, or associated audio services that are complete services, encoded at a bit rate up to and including 384 kbps. Transmissions may contain single-channel associated audio services intended to be simultaneously decoded along with a main service encoded at a bit rate up to and including 128 kbps. Transmissions may contain dual-channel dialogue associated services intended to be simultaneously decoded along with a main service encoded at a bit rate up to and including 192 kbps. A further limitation on transmissions is that the combined bit rate of a main and an associated service that are intended to be simultaneously reproduced is less than or equal to 512 kbps.

ANCILLARY DATA SERVICES

Unlike the NTSC system, and the monochrome TV standard before it, the digital television system delivers digital data that represents video and audio to receivers that the receiver converts back into sound. But any type of information can be delivered in the form of digital data, and the digital television system takes advantage of this characteristic, affording broadcasters the opportunity to augment the basic television services with ancillary digital data services. The flexibility of the MPEG standard allows new digital services to be easily introduced at any time in such a manner that it is compatible with the existing system.

This chapter will describe ancillary services that are specified as part of the DTV standard from the outset, and will also describe the provision that exists for new services that may be added in the future. The ancillary services that are part of the DTV standard are the ones that provide for the transmission of program subtitles (PS), emergency messages (EM) and program guide information.

Text-Based Ancillary Services

The DTV standard includes provision for two text-based services that augment the basic television service: program subtitles and emergency messages. The program subtitles are analogous to the closed caption service transmitted on line 21 of the NTSC signal, and emergency messages, which are mixed into the baseband video signal in the NTSC system.

THE PROGRAM GUIDE

Broadcasters have the option to broadcast an interactive program guide database as part of the transport stream. This database would contain information about the programs being transmitted at that moment, as well as information about programs that would be transmitted in the future. It would also include time and control information. ATV receivers that are equipped with the necessary circuitry would be able to present an on-screen grid of program information.

3

SERVICE MULTIPLEX AND TRANSPORT SYSTEMS

In order for the program material of an ATV program to be used to modulate an RF carrier to be transmitted via either a terrestrial broadcast channel or a cable channel, the data that makes up the bit streams that represent the video and audio portions of the program must be formatted and multiplexed. The portion of the transmission system that accomplishes these actions, and the portion of the receiver that reverses these actions, make up what is referred to as the *transport* subsystem. This chapter will describe in a general way what takes place at both ends of the transport subsystem.

The transport system is based on fixed-length transport stream packetization, as defined by MPEG. This type of transport layer fits well with the characteristics of both terrestrial broadcast and cable. The use of moderately long, fixed-length packets fits in with the needs and techniques for error protection in both types of environments. This approach is also flexible enough so that it can accommodate the initial needs ATV to multiplex video, audio, and data, while providing the capability to add additional services in the future that are compatible with the existing system. Another advantage of basing the transport layer on MPEG-2 is that this assures that the system will be compatible with other media and standards.

DYNAMIC CAPACITY ALLOCATION

The use of fixed-length packets affords total flexibility to allocate channel capacity among video, audio, and auxiliary data services. The use of a packed identifier (PID) in the packet header to identify that portion of the bit stream makes it possible to have a combination of video, audio and auxiliary data that is not only flexible, but it is also not necessary to specify it in advance. The channel can be reallocated for varying periods of time to allow for delivery of data.

Here's an example of the usefulness of such a scheme. Let's say that the broadcast station was sending a program for a short period of time that consisted of a still picture, and no audio, for example a company's logo, at the end of a commercial. There would be a lot of unused bandwidth. During such a period, the broadcaster could transmit program-related computer software to a "smart recover."

Scalability

The transport format is said to be "scalable." That means that if a channel is available that has greater bandwidth, the broadcaster can send more information down that larger channel. For example, the broadcaster can add more elementary streams at the input of the multi-plexer. Another way to send more information through a larger channel is to multiplex those additional bit streams with the original bit stream at the second multiplexing stage. Scalability is useful in the case of network distribution, but it is particularly valuable for cable systems that will be capable of delivering higher data rate within a single 6 mHz channel.

Extensibility

No matter how carefully a system is planned, something else unforeseen always seems to come along at a later time that would make a useful addition to the system. In some cases, if provision is not made for such services, it may not be possible to add it at all. Or it may take a costly change or addition to the system to accommodate it. For example, what would happen if, at some time in the future, broadcasters want to add 1000-line progressive formats or 3D-HDTV?

The ATV system has made provision for such features. The transport architecture provides open-ended extensibility of services. If some new service becomes available, new elementary bit streams could be handled at the transport layer without making any hardware modifications. The broadcaster would simply assign new packet IDs at the transmitter. New TV sets that were designed to offer these new services would be designed to filter on these new PIDs in the bit stream. And the addition of these new bit streams with their new PIDs won't cause older TV sets that aren't equipped to handle them obsolete. Decoders in older, existing, TVs will simply ignore them. In other words, this insures backward compatibility.

Robustness

In the parlance of HDTV, or even computers and their software, the term robustness implies the ability to handle errors without causing some kind of serious, or even total, degradation of the system performance. For example, in computers, some software operating systems are called robust because they are designed so that the computer won't crash and have to be restarted even if some kind of error occurs. In the case of HDTV, the system of using the fixed-length packetization approach is said to be robust because if some kind or problem occurs that interrupts transmission, the fixed-length packet can form the basis for handling errors that occur during transmission so that the signal can be quickly be reacquired at the receiver.

At the receiver, error correction and detection processing comes before packet demultiplexing. This process may be synchronized to the packet structure, so that the decoder deals with units of packets when handling data loss that occurs as a result of transmission impairments. Here's what happens. After the receiver detects the fact that a transmission error has occurred, it starts to look for a good packet. When the first good packet is received after the disruption, the receiver uses that to recover the data bit stream. Recovery of synchronization is also aided by the header information contained in the transport packet.

Cost Considerations

The use of fixed-length packets in the transport system helps keep TV receiver costs down, because it allows set designers to use simple decoder bit stream demultiplex circuit and software designs that are suitable for high speed operation. The decoder doesn't need to be programmed to have the information describing how the signal was multiplexed, or the bit-rate characteristics that exist at the transmitting end in order to extract elementary bit streams at the demultiplexer. The receiver only has to know the identity of the packet, which is transmitted within the packet header at fixed and known locations in the bit stream.

MPEG-2 Compatibility

The transport system of the ATV system is based on the MPEG-2 system specification. This specification is very broad, and therefore can accommodate a number of different approaches to storage of program data as well as transmission of program data. If the ATV system were to be designed to be able to accommodate everything that is possible with MPEG-2, the system would be less efficient than it needs to be. Therefore, the ATV system has been designed to use a limited subset of what the MPEG-2 system specification allows.

Besides allowing broadcasters flexibility in storage and transmission of program information, use of the MPEG-2 standard for broadcast makes HDTV programs compatible with other MPEG-2 applications, such as storage of compressed bit streams, computer networking, and systems that deliver non-HDTV data. This ability of data to be handled by any of a number of computer-based systems has been termed "interoperability." This compatibility, or interoperability, implies that DTV transport bit streams may be handled directly in these other applications.

Because, as we said at the outset, the ATV system will not exercise all of the capabilities defined in the MPEG-2 standard, the decoder in a digital television receiver will not have to be fully compliant with the MPEG-2 system specification. It will, in other words, not be required to be able to decode any arbitrary MPEG-2 bit stream.

THE TRANSPORT SUBSYSTEM

For an idea of the complete transport subsystem as a single HDTV program is being broadcast from the transmitter and received at the receiver, see *Figure 3-1*. The transport system is said to "reside" between the application (audio or video, for example) and the transmission

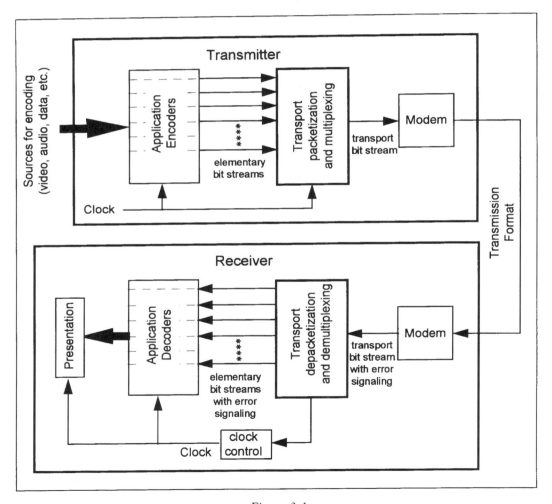

Figure 3-1

subsystem. That is, the functions of the transport subsystem at the transmitting end operate on the audio and video program material in such a way as to prepare them to be transmitted. At the receiving end, the transport subsystem operates on the received program in such a way as to convert it to the audio and video program information.

At its lowest layer, the encoder transport subsystem formats the encoded bits that comprise the program material and multiplexes the components of the program in preparation for transmission. The portion of the transport subsystem that exists in the receiver in the viewer's living room, the transport subsystem recovers the bit streams for the individual application (audio, video) decoders, and for the corresponding error signaling.

At a higher layer, the transport system can accommodate a number of different programs at the same time. In this case, there would be an additional system level multiplexing stage before the modem in the transmitter, and an additional system level demultiplexing stage after the modem in the receiver.

The transport subsystem incorporates other higher level capabilities that have to do with identification of the program material (applications) and as shown in the figure, synchronization of the receiver.

We mentioned earlier that the data transport system is based on the use of fixed-length packets, each of which is identified by a header. Each header identifies the packet as a portion of a particular application bit stream (also called an elementary bit stream). The remainder of the data in the packet is a portion of the audio or video (or it could be other types of data) that forms part of the program: in ATV parlance, the "payload."

Types of data (applications) supported include video, audio, data, program, and system control information. The elementary bit streams for video and audio are divided into a variable-length packet structure called the packetized elementary stream (PES) before being processed by the transport system. The activities that take place in the PES layer allows the receiver to identify the portions of the individual applications so that they (audio and video) can be decoded and presented in synchronization.

Elementary bit streams that have a common time base (such as the audio and video portions of a television presentation) are multiplexed along with a control data stream into *programs*. Here's an important distinction in terminology: in the digital television system, a *program* is analogous to a *channel* in the NTSC system, in that it contains all of the video, audio and other information necessary to make up a complete television program. These programs are then multiplexed with an overall system control data stream and then multiplexed asynchronously to form a multiplexed system.

At this level, the transport subsystem provides for a great deal of flexibility.

- It permits programs to be defined as any combination of elementary bit streams. To be more specific, the same elementary bit stream can be present in more than one program. As an example, two different video bit streams can be identified with a single audio bit stream. Another possibility is that a program can be formed by combining a basic elementary bit stream with a supplementary elementary bit stream, as for example, bit streams for scalable decoders. As yet another possibility, programs can be programmed for specific conditions. As an example of this, a broadcaster may decide on a regional basis a language to include as a secondary audio bit stream.

- Flexibility at the system's layer allows different programs to be multiplexed into the system as deemed useful by the broadcaster, and allows the system to be reconfigured easily when that is necessary. The procedure for accessing a separate program from within a system is also simple.

The transport system provides other features that are useful for normal decoder operation as well as for special features that are required in broadcast and cable operations, including decoder synchronization, conditional access, and local program insertion.

Because of the way the transport bit stream is defined, the included provisions for an entity, such as a television production company or a broadcast station, to create a program, store it, and transmit it at the scheduled time. The programs are stored in the same form as that in which they're transmitted: as transport bit streams.

And here's an interesting thought: HDTV receivers may be designed with the ability to store the incoming program bit streams, and play them back at the pleasure of the viewer. This capability may include playing the program in slow motion, stopping the motion, or showing it backward. The transport bit stream format also contains the information, or "hooks," that will allow these "trick" modes of play similar to those possible with VCRs. It's important to be aware that recording and playback of digitally compressed video bit streams is much different from recording and playback of analog television material such as NTSC.

INTEROPERABILITY

We've touched on this on a few occasions before in this book, but let's state it explicitly here. Companies involved in all aspects of the delivery and manipulation of information feel that it's a good idea to design systems such as HDTV, personal computers, video games, and anything else you can think of, such that any one of those products will be able to operate using data created on any of the other systems. A term that has been developed in the consumer electronics business to describe the idea that televisions, stereo systems, VCRs, camcorders, personal computers, communications systems, and more are all tending to become more and more interrelated, even more or less separate aspects of the same technology, is *convergence*. A term that has been coined to describe the design of these various systems so that they can all use the data created for use on any one of them is called *interoperability*. Interoperability will foster convergence.

Bit stream interoperability at the transport level is an important feature of the digital television system. The question of interoperability is one that goes in both directions. On the one hand, the question is whether the transport bit stream can be carried on other communications systems. On the other hand, the question is whether the digital television system can carry bit streams generated on other communication systems.

In general, says the ATSC documentation, there is nothing that prevents the transmission of a bit stream as the payload of a different transmission system. It may be easier to perform this function on some systems than on others, but it is always possible. Because asynchronous transfer mode (ATM) is expected to form the basis of future broadband communications networks, it is especially important that bit streams be able to operate on these systems. Because of the great importance attached to this particular interoperability, it has been specifically addressed in the design of the digital television system.

PACKETIZATION AND FUNCTIONALITY

The transport bit stream consists of fixed-length packets. Each packet contains a header field that consists of a fixed component and a variable component, and a "payload" of data (*Figure 3-2*).

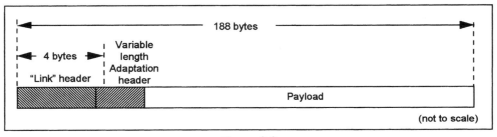

Figure 3-2

As shown in the figure, each packet consists of 188 bytes, and is created in accordance with the MPEG-2 transport system specification syntax and semantics. The size of the packet was chosen based on several factors. For starters, it was deemed important that the packet be large enough so that the amount of data carried in the transport header would not be a significant portion of the total data in the packet, compared to the payload data.

Conversely, packets should not be so large that it becomes likely that a packet error will occur under normal conditions as a result of inefficient error correction. The designers of the system also wanted to choose the lengths of the packets so that they were consistent with the block sizes of typical, block oriented, error correction methods. This would allow the system to synchronize them to error correction blocks, so that the physical layer of the system can aid the packet-level synchronization process in the decoder.

Another reason that this particular packet length was selected is so that this system would be compatible (interoperable) with the ATM format. The general idea is to transmit a single MPEG-2 transport in four ATM cells. This method of transmission can be performed in any one of a number of ways.

The packet header identifies what is contained in the packet. The packet header structure is "layered," and may be described as a combination of a fixed-length "link" layer, and a variable length adaptation layer. In the digital television system, this approach of using a link layer and an adaptation layer are used directly for the terrestrial broadcast system on which the MPEG-2 transport bit stream is broadcast.

If these same packets were being transmitted via a different communications system, say ATM, the MPGE-2 packet headers would not be used in any way to control or otherwise affect the transmission. They would simply be carried as part of the payload and would continue to be served to identify the contents of the data stream.

The Packet Header Link Layer

The link layer contains a four-byte header field. We'll talk about it in greater detail later. The link layer provides for the following four functions:

• Packet synchronization

• Packet identification

• Error handling

• Conditional access

Here's what those terms mean.

Packet Synchronization

The first byte in the packet header, called the sync-byte, performs the synchronization function. The sync-byte has the same fixed, preassigned, value for all MPEG-2 bit streams. Decoders in some HDTV receiving sets perform packet synchronization in the circuits of the communication link, which comes before the packet demultiplexing stage (which means that the synchronization information isn't available to the receiving set yet). In this case, the sync-byte field may be used for verification of the packet synchronization function.

In some other decoders in other HDTV receivers, the sync-byte field may be used as the primary source of information for synchronizing the received packets. The ATV standard doesn't specify how the set manufacturer gets this function done, it merely provides the information in the bit stream to make it possible to do it.

Packet Identification

When a number of bit streams are going to be mixed together in some way (multiplexed), it's very important that the individual bit streams carry some kind of identification so that they can be sorted out (demultiplexed). The information in the header that carries this information is the PID (packet identification) field. This field identifies the packet as belonging to a particular elementary or control bit stream. The location of this field in the packet header is always the same. Because the location of this field is fixed, once packet synchronization is achieved it becomes simple to extract packets that contain portions of an elementary bit stream by filtering packets based on this PID. The fixed packet length makes it simple to filter and demultiplex them, and so is suitable for high-speed transmission systems.

Error Handling

Error handling in the ATV system is conceptually simple, and elegant, and apparently works very well. In the packet header there is a field called the continuity-counter field. At the transmitter, the value in this field, for all packets in succession that have the same PID and that carry a data payload, is cycled from 0 to 15, then starts over at 0. It should be noted that the ATV system transport protocol allows packets to be defined that don't have a payload.

Under normal conditions, if the home HDTV receiver receives packets for which the values in the continuity-counter field are not in proper order, the set knows that data has been lost in transmission. As a concrete example, let's say that the previous packet that the HDTV set received had a value in the continuity-counter field of 05, and the current value is 12, packets 6 through 11 were not received. In this case, the transport processor in the receiving set would signal the decoder for that particular elementary stream that data had been lost. The way in which any HDTV set manufacturer handles this signaling is entirely up to that manufacturer: it is not part of the ATV standard.

Certain types of information (such as adaptation headers, time stamps, and program maps) are of vital importance to the smooth and continuous operation of the system. The transport system has a means to make this information less susceptible to events that might disrupt the program channel. It does this by duplicating that information at the encoder and sending it more than once. When these duplicate packets reach the receiving HDTV set, if no error occurred, and the original packet arrived properly, they will be ignored. If, however, the original was in error, or didn't make it to the receiver because of a disturbance in the transmission medium, the duplicate will be used instead.

Conditional Access

Of course, there are times when certain broadcasters want to scramble their signal, as for a pay-per-view event. The transport format allows for scrambling of data in the packets. Each elementary bit stream in the system can be scrambled independently. The digital television standard specifies how the signal is to be scrambled, but it doesn't specify what the descrambling key will be, or how the decoder in the receiving set will obtain it. The key must be delivered to the decoder within a time interval of its use.

There is a "private" data capacity at several locations within the transport stream where this key data might be carried. Here are a couple of possibilities:

• A separate private data bit stream with its own PID.

• A private field within an adaptation header carried by the PID of the signal being scrambled.

The security of the conditional access system is ensured by encrypting the descrambling key when sending it to the receiver and by updating the key frequently. The digital television system standard doesn't add constraints on the number of keys and the rate at which they may be changed.

Information in the link header of a transport packet indicates whether the payload portion of the packet is scrambled. If it is, that information flags the key to be used for descrambling. The header information in the packet is always transmitted unscrambled. The amount of data in a packet that is to be scrambled varies, depending on length of the adaptation header. It should be noted that some padding of the adaptation field may be necessary for certain block mode algorithms.

Within the MPEG-2 standard, the transport definition allows the transmitting end to scramble at two levels: within the structure of the PES packet, and at the transport layer. Scrambling at the PES packet layer is primarily useful in the program stream (which is not supported in the digital television standard), where there is no protocol layer similar to the transport layer to enable this function. In the digital television system, scrambling will occur only at the transport layer.

The Adaptation Layer

The adaptation layer in the MPEG-2 packet is a variable-length field. Its presence is flagged in the link level section of the header. So, what are these headers for? They're necessary for the decoding of the elementary bit stream that is extracted using the link level functions. Some of the functions of this layer that are important to the operation of the digital television system follow.

SYNCHRONIZATION AND TIMING

The idea of synchronization in the digital television system is very different from that in an analog system, such as the NTSC system. In the NTSC system, the receiver can use the synchronizing information in the received signal to derive a clock that provides timing for displaying the picture and sound information. In a digital compressed system, such as digital television, the amount of data generated for each picture is variable, depending on the picture coding approach and the complexity of the data. In short, there is really no natural concept of synchronism between transmission of the program and display of the program in a compressed digital television bit stream, compared to the synchronism that we have become used to in NTSC.

But synchronism, in the context of the digital television system, is important. Received data is expected to be processed at a particular rate. That rate matches the rate at which the data is generated and transmitted. Loss of synchronization leads to serious problems: buffer overflow or underflow at the decoder, which culminates in loss of synchronization in presentation or display.

So how does the system keep loss of synchronization from happening? The answer is that the transmitting station sends timing information in the adaptation headers of selected packets to serve as a reference for timing comparison at the decoder. They do this by transmitting a sample of a 27 MHz clock in the program_clock_reference (PCR) field. This sample indicates the expected time at the completion of the reading of that field from the bit stream at the transport decoder. The phase of the local clock running at the decoder is compared to the PCR value in the bit stream at the instant at which it is obtained, to determine whether the decoding process is synchronized.

In general, the PCR from the bit stream doesn't directly change the phase of the clock in the receiving set, but merely serves as an input to change the clock rate. There are exceptions to this statement: during channel change and during insertion of local programming. Remember, the nominal clock rate in the decoder system is 27 MHz. Here's something to keep in mind: the digital television standard specifies the means of transmitting synchronizing information to a receiver. It doesn't specify how the manufacturer of the HDTV set will effect recovery of synchronization.

Something else to keep in mind is that the audio and video sample clocks in the receiver decoder system are locked to the system clock derived from the PCR values. This simplifies the design of the television receiver as regards the number of local oscillators that are needed to drive the complete decoding process. It has other advantages, as well, such as rapid sync acquisition.

Random Entry into the Compressed Bit Stream

The digital television system has to make it possible for a receiving set to randomly enter the application bit streams, such as audio and video, so that the receiver can pick up (acquire) the signal when the viewer turns on the set, or changes channels (Oops, that's an NTSC term. We meant changes *programs*). It's possible to enter into the elementary bit stream only if the coding for the elementary bit stream for the application supports it directly. For example, the video bit stream makes provision for random entry through the concept of interframes (or I-frames, remember those) that are coded without any prediction and, therefore, can be decoded with no other information available. A good point for the receiver to start receiving the program (randomly enter a video bit stream) would be at the beginning of the video sequence header information that precedes the data in an I-frame. More generally, any time a digital television receiver begins receiving a program, either when the set is first turned on, or when the channel is changed, or after a disruption of the signal (called random entry by the ATV standard) it should coincide with the start of PES packets when they are used: for example, for video and audio.

The transmitted signal contains the information, at the transport layer, that the receiver needs to tell it when to jump into the data stream, so to speak. There's a flag in the adaptation header of each packet that indicates whether that packet contains a random access point for the elementary bit stream. In addition, if the packet is a random access point, the data payload starts with the data that forms the random access point of entry into the elementary bit stream itself.

Because the digital television system works in this manner, when the viewer turns the set on, or changes channels, or when there has been a disruption, and the receiver is looking for a resynchronization point in the transport bit stream, the receiver can just ignore packets, until it sees one with this flag. It also makes it easier for the receiving set to find an entry point in the elementary bit stream once transport level synchronization has taken place.

The whole idea of this scheme is to have frequent random entry points into the HDTV programs so that the viewer can change channels as rapidly as she wishes to.

Inserting Local Programs

The transport system makes it possible for broadcasters to insert local programs and commercials by including provisions such as flags and features especially for this purpose in the transport packed adaptation header. This important issue has been covered at great length by the MPEG. The MPEG-2 *syntax* (syntax as regards data and programming describes the rules governing the employment of the programming language) not only allows local programs to be inserted, but it allows its performance to improve as techniques and equipment are developed around these syntax tools.

There are a lot of other considerations that concern switching and insertion of programs at the broadcast end, but since it doesn't really affect the receiving set one way or another (the set will decode and display whatever program the transmitter sends) we won't go into any more detail about it here.

HIGHER-LEVEL MULTIPLEXING

The question of multiplexing is a different story, however. Multiplexed programs will have to be demultiplexed at the receiver, so anyone wanting to know about HDTV will need to know what's going on with multiplexing. The approach to multiplexing in digital television includes multiplexing at two different layers. In the first layer, program transport streams are formed by multiplexing one or more elementary bit streams at the transport layer, and in the second layer, more than one transport streams are combined, using asynchronous packet multiplexing, to form the overall signal. The functional layer in the system that contains both the program and system level information is called Program Specific Information (PSI).

Single Program Transport Multiplex

To try to keep confusion about terminology to a minimum, and in a system as complex as digital TV confusion is inevitable, let's look at a couple of definitions. In digital TV, the term *program* means roughly the same thing as does the term *channel* in the NTSC system. The MPEG defines the term *program* stream as a particular bit stream format. This term is not used in the digital television standard. In the digital television standard, the term *program transport stream* describes a transport bit stream that has been generated for a program. If you think about it long enough, it might make sense. I'm still working on it.

A program transport stream is formed by multiplexing individual elementary bit streams (these bit streams may or may not be packetized into packetized elementary streams [PES] that have the same time base). As the elementary streams are multiplexed, they are formed into transport packets, and a control bit stream that describes the program is added. This control bit stream, which is also sometimes called the elementary stream map (as in *Figure 3-3*) is also formed into transport packets. The elementary bit streams and the control bit stream are identified by their unique PIDs in the link header field. The way this multiplex function comes together is shown in *Figure 3-3*.

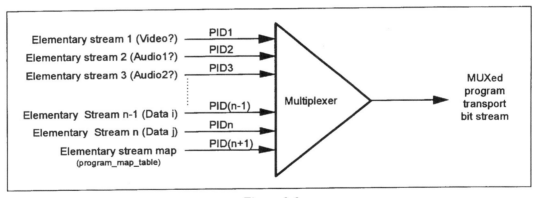

Figure 3-3

The control bit stream contains the program_map_table that describes the elementary stream map. The program_map_table includes a great deal of information:

• Information about the PIDs of the transport streams that make up the program.

• The identification of the applications (for example, audio and video) that are being transmitted on these bit streams.

• The relationships among these bit streams, and other pertinent information, if any.

The details of the program_map_table syntax and details as to what each syntax element accomplishes will be covered in a later section. A bit stream carrying a program_map_table is identified at the system layers, which will be described next.

The syntax of the transport system allows a program to contain a large number of elementary bit streams. There is no restriction on the types of applications that are required in any program. Here's an example: a program transport stream doesn't have to contain a single video or audio bit stream. It could be carrying a data "program" instead. However, another possibility is that a program transport might contain a number of related video and audio bit streams, as long as they all have the same time base.

The data applications that the transport system carries are flexible. The only requirement is that there has to be an appropriate stream_type ID assignment so that the decoder at the receiving unit can recognize the application that corresponds to the bit stream.

System Multiplex

A number of different program transport streams can be multiplexed together. This is known as system multiplex. In order for this to succeed, the broadcaster has to define a system level control bit stream with PID=0, that is transmitted in addition to the transport bit streams that define the individual programs. This bit stream carries an entity known as the program_association_table that maps program identities to their program transport streams. The identification of the program is represented by a number in the program_association_table. At the risk of repeating this once too often, a program in the digital television system corresponds to what has been traditionally called a channel in the NTSC system, for example, PBS, WBZ, WGN, etc. The map indicates the PID of the bit stream containing the program_map_table for a program.

Here's what happens at the HDTV receiver. In essence, identifying a program and its contents takes two steps:

- First, the receiver uses the program_association_table in the PID=0 bit stream to identify the PID of the bit stream carrying the program_map_table for the program.

- Next, the receiver obtains the PIDs of the elementary bit streams that make up the program from the appropriate program_map_table.

After these two steps have been accomplished, the filters at a demultiplexer can be set to receive the transport bit streams that correspond to the program of interest. Another use of this information is to create a program guide at the receiver.

Figure 3-4 shows the system layer of multiplexing. Unfortunately, if something weren't done to prevent it's happening, during the process of system level multiplexing, it would be possible that PIDs on different program streams would be identical at the input. But PIDs for different bit streams, of course, have to be unique. This problem can be solved at the multiplexing stage, by modifying some of the PIDs just before the multiplexing operation. Any such changes have to be recorded in both the program_association_table and the program_map_table. The ability to perform this PID reassignment function through hardware (in other words, circuitry) in real time is aided by the fact that this process is synchronous at the packet clock rate. Another way to accomplish this reassignment of PID numbers would be to make sure at the outset that the PIDs being used in the programs that comprise the system are unique. When the broadcaster is dealing with stored bit streams, that is not always possible.

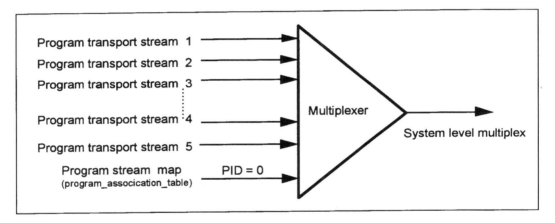

Figure 3-4

Still higher levels of multiplexing can be accomplished. In the vernacular of the ATV standard "...the architecture of the bit stream is scalable." A number of system-level bit streams can be multiplexed on a broader bandwidth channel by extracting the program_association_tables from each system multiplexed bit stream and reconstructing a new PID=0 bit stream. Once again, the PIDs may have to be reassigned in this case.

The ATV standard doesn't include any kind of standard for the multiplexer, or how it is to be used. The Advanced Television System Committee leaves it up to individual designers as to how to accomplish multiplexing. At the receiving end, demultiplexing is a matter of filtering, so the transport demultiplexer will demultiplex any digital TV bit stream, no matter how the transmitting end did the multiplexing.

Figure 3-5 is an illustration of how the elementary bit streams will be extracted at the receiver. It also presents one method of actually performing the demultiplexing process. However, this may not be the most efficient way of going about this process.

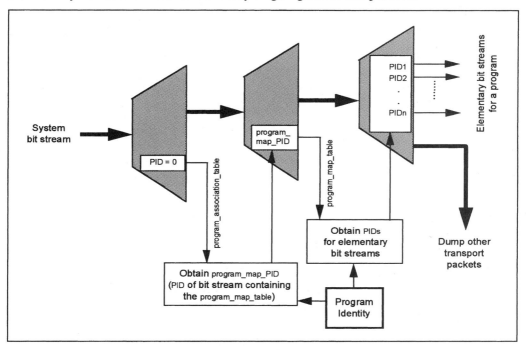

Figure 3-5

In an actual HDTV set, it would be possible to extract both the program_association_table and the program_map_table control bit streams using the same demultiplexer hardware. The drawing also represents the least amount of functionality required to extract any application bit stream (including any private bit streams if there are any).

The demultiplexing is just the first step in the process of receiving an HDTV program. Once the packets are obtained for each elementary bit stream in the program, the receiver will have to process them further to accomplish the following:

• Obtain the random entry points for each component bit stream,

• Achieve decoder system clock synchronization,

• Obtain presentation (or decoding) synchronization,

before the receiver reaches normal operating conditions for receiving a program.

THE TRANSPORT FORMAT AND PROTOCOL

This section is important to anyone who wants to understand what HDTV is all about. It discusses the definition of the syntax elements for the transport layer bit stream. And here's the key phrase from the **Guide to the Use of the ATSC Digital Television Standard**: "All syntax elements need to be recognized at some level in a digital television receiver. Most syntax elements trigger a response in the transport decoder." The guide goes on to state that some of the syntax elements are present so that the digital TV signal is "interoperable" with MPEG-2.

Link-Level Header

Figure 3-6 is an illustration of the link level header, showing what each bit, or group of bits does. When the HDTV signal is transmitted over a standard terrestrial broadcast channel, the functions shown may not all be used. The functions that are not used for terrestrial broadcast are there for transmitting the same bit stream over other links: cable, computer networks, etc. In other words, those functions provide interoperability features.

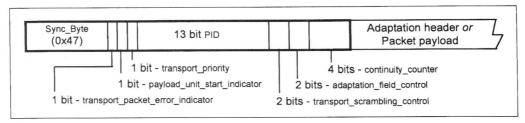

Figure 3-6

Again, referring to *Figure 3-6*, the bit marked transport_packet_error_indicator is there to signal errors that have occurred between the modem and the transport demultiplexer. If this bit is set, the receiving set will not use the payload of the packet. The concept of prioritized delivery of data is not applicable in the digital television system, so the receiver simply ignores the bit (called a flag) marked transport_priority. Since the program from the transmitter might feed a prioritized delivery system, as well as HDTV receiving sets, this flag might be set. The payload_unit_start_indicator is used to let the receiver know that the payload of the packet contains the start of an entity that contains the payload data. In the case of audio and video, this is the PES packet. For PSI data, the payload data is the start of a table.

The adaption_field control bits tell the receiver whether or not an adaptation header field is present, and if so, whether or not there is also a payload in the same packet. In the case of transport packets containing PSI information, adaptation headers are not allowed by the standard, except for signalling with the discontinuity_indicator that the version_number may be discontinuous.

The transport_scrambling_control indicators can take on any of four values: 00, 01, 10, 11. Have a look at *Table 3-1* for an explanation of what those values tell the receiving HDTV set.

transport_ scrambling_ control	Function
00	packet payload not scrambled
01	not scrambled, state may be used as a flag for private use defined by the service provider.
10	packet payload scrambled with "even" key
11	packet payload scrambled with "odd" key

Table 3-1

Adaptation-Level Headers

If the transport packet contains an adaptation field, the adaptation_field_control bits of the link level header will reflect that, as mentioned just above. The information in the adaptation header itself is useful for higher level decoding functions. The format of the header involves flags to indicate the presence of the particular extensions to the field.

The header starts with a fixed-length 2-byte component that is always present (if an adaptation header is transmitted). The format for this fixed-length component of the adaptation header is shown in *Figure 3-7*.

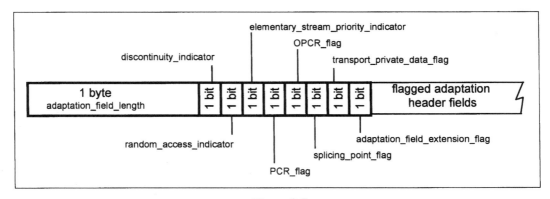

Figure 3-7

The byte marked adaptation_field_length specifies the number of bytes that follow it in the adaptation header. The adaptation header may include "stuffing" bytes after the last adaptation header component field. These stuffing bytes have a value of 0xFF. They are not interpreted at the decoder. If the adaptation header includes stuffing bytes, the adaptation_field_length includes them in the count. In some instances, the value in the adaptation_field_length field can also be used by the decoder to skip over the entire adaptation header, and to directly advance to the data payload in the packet.

The state of the last five single bit flags will indicate whether additional adaptation header fields are present. A value of 0 will indicate that the corresponding field is absent, whereas a 1 will indicate that the corresponding field is present. Regardless of the value of the three flags that come before these five fields, there will not be an extension of the adaptation header. The field identified as elementary_stream_priority_indicator is not used in the digital TV system. If this field is set at the transmitting end, the decoder in the receiver will simply ignore it.

As stated earlier, the state of the flags in the fixed-length portion of the adaptation header, as shown in *Figure 3-7*, will signal to the receiving digital television set if the corresponding components of the adaptation header will be present. The order in which these components appear is the same as the order of the flags. Depending on the type of adaptation header information being conveyed, the data in these fields may be either fixed length or variable length. Details on the contents of these fields follow.

THE PCR AND OPCR FIELDS

The purpose of the PCR (program clock reference) was described in some detail earlier. This section will also address the concept of the original program clock reference (OPCR).

The PCR is used to synchronize the system decoding process. Under some circumstances, the PCR field may have to be modified during transmission. The PCR will be transmitted at least every 100 milliseconds (one-tenth of a second).

The original program clock reference (OPCR) indicates the intended time of arrival of the last byte of the original_program_clock_reference_base at the decoder for a single program. This field remains unchanged during transmission, and may be used for recording and playback of single programs. The receiver does not use the OPCR during playback.

The Splice Countdown Field

The splice_countdown field contains information that indicates how many packets with the same PID as the current packet remain in the bit stream until there is a splicing point packet. This field signals to the receiver the point at which a switch from one program segment to another program segment may occur. Once the receiver has begun to receive, decode, and display the inserted program segment, this field tells the receiver when the inserted segment will end and the original program will resume.

The splicing point packet is defined as the packet that contains a point in the elementary bit stream from which point onward data can be removed and replaced by another bit stream so that the resulting bit stream is valid, according to the MPEG-2 standard.

PSIs and the Pointer Field

Remember that PSI is the abbreviation for program specific information. As we've implied in several sections in this book, there's a lot going on in the origination, digitization, broadcast and reception of the digital television signal. And as they say about sports, "you can't tell the players without a program." Well, it's the same in DTV. With all the multiplexed bitstreams coming into the receiver, there has to be some kind of information to tell the receiver what's what.

That's where the program_association_table and the program_map_tables come in. These tables, which constitute a part of the PSI layer, carry information that describes the way in which a multiplexed bit stream is organized. PSI tables are transmitted in the appropriate bit stream in sequence. There are no gaps between the tables. This means that tables don't necessarily begin at the beginning of a transport packet. Some method must be provided, therefore, to indicated where these tables begin in the bit stream in order to decode specific tables. That method is called the pointer_field.

If the pointer_field is present in a packet, it is the first byte of the payload of the packet, and follows after the link and adaptation headers. If a PSI table begins in the packet, this is indicated at the link level by setting the payload_unit_start_indicator to "1," and the pointer_field will be present in a packet. The pointer_field indicates the number of bytes that follow it before the start of a PSI table. Here's the most basic example: if the pointer_field has a value of 0x00, this indicates that a new PSI table begins immediately following it.

The Program_Association_Table

It was mentioned earlier that the program_association_table describes how program numbers that are associated with programs (PBS, C-SPAN, etc.) map on to bit streams containing the program_map_tables for the respective programs, and that the program_association_table is transmitted as the payload of the bit stream with PID=0. The program_association_table may be transmitted as multiple program_association_segments. In this situation, each segment will have a maximum length of 1024 bytes. The transport decoder at the receiving DTV set can extract individual table segments from the bit stream in whatever order it chooses.

The component of the table that is variable in length consists of a number of fixed-length entities corresponding to each program. The number of such entities is contained in the byte designated program_count. This table component will also contain stuffing_bytes to make up the program_association_segment_length.

The program identity '0' is reserved for the network_PID, that is, the PID of the bit stream that carries information about the configuration of the overall system. The network PID is available for use by broadcasters to provide information about the programs being transmitted currently, as well as for programs that will be transmitted at a future time. For all other program identities, the program_map_PID is the PID of the bit stream that contains the program_map_table for the program that it's being transmitted with.

The Program_Map_Table

In the previous section, it was stated that the program_map_table is transmitted as the payload of the bit stream with PID=program_map_PID (indicated in the program_association_table). The program_map_table carries information about the applications that make up programs, such as audio, video, etc. The format of the first eight bytes is the same as that of the program_association_table, and performs the same function. These are made similar in format to make possible simple software decoding of headers.

Descriptors

To describe certain characteristics of the program or the elementary bit stream, information, called descriptors, are transmitted as part of the transmission. These are referred to as the program_descriptor and the elementary_stream_descriptor fields. Each descriptor can consist of a number of descriptor field elements transmitted sequentially.

But as with anything in this system, there are two things that need to be considered in using descriptors.

For one thing, there has to be some way to indicate that descriptors are present. In the PSI tables that have been described, the presence of descriptors is signalled by the length field that comes just ahead of the descriptor. A value of zero in this field indicates that there is no descriptor present. Second, the descriptor itself has to be identified. This identification takes place in the descriptor header. This identification consists of a one byte descriptor_tag field, followed by a one byte descriptor_length field. The descriptor_length field specifies the number of bytes in the descriptor following the descriptor_length field. The set of valid descriptor_tags is the same as that defined by MPGE-2.

THE PSI PARADIGMS AND CONSTRAINTS

People who watch NTSC programs today, and who doesn't, expect to be able to flip through channels pretty quickly. They're not going to be very happy with digital television if they can't do the same thing, or pretty close, anyway. So for digital TV broadcasters and set manufacturers, the time to acquire programs is a very important concern. Rapid access to programs will be assured by defining a repetition rate for transmission of different PSI bit stream elements and assigning PID values to bit streams that are related to a given program. The program paradigms (nobody really knows what *paradigm* means, but we're going to try to define it in the next section) are such that recognition of the paradigm helps speed up the acquisition of programs in the receiver. A receiver that is not aware of the program paradigms will be able to acquire the program, but the viewer may have to slow down his channel surfing speed.

The Program Paradigms

OK, here we go with an explanation of *paradigm*. The dictionary defines paradigm as an example, or pattern. Some of the modern usages of the word suggest that a paradigm is sort of a whole way of looking at something, and it gets pretty philosophical, talking about paradigm shifts, and so forth, but there's the actual definition of paradigm. *A pattern.*

And in terms of digital television, that's pretty much what paradigm means: a pattern of data that is incorporated into the television that helps it recognize when a given program is being received. The program paradigms provide a simple mechanism for quickly recognizing the programs of interest and further identifying the transport bit streams containing elements of the program, thus enabling quick acquisition. The general idea is to select PIDs for the transport bit stream that are related to a program number. The program number within a single transport stream serves as the complete identification of the program.

Television programs can only be assigned program numbers from 1 to 255. Nontelevision programs are not allowed to use these numbers. This system allows the receiver to quickly filter television programs by looking at the most significant bit (MSB) of the program number. This process is further aided by specifying television programs to be described only within section 0 of the program_association_table.

Here's another definition: base_PID = program number << 4 (pseudo C code), where program number refers to each program within one transport multiplex and corresponds to the 16-bit program_number identified in PAT and PMT.

The paradigm to identify the transport bit streams containing certain elements of the program is defined as shown in *Table 3-2*. *Table 3-3* illustrates an example for program 52.

Name	PID Definition	Description
PMT_PID	base_PID+0x0000	PID for the bit stream containing the program_map_table for the program.
Video_PID	base_PID+0x0001	PID for the bit stream containing the video for the program.
PCR_PID	base_PID+0x0001	Implies the video bit stream also carries the PCR values for the program
Audio_PID	base_PID+0x0004	PID for the bit stream containing the primary audio for the program. The primary audio shall be a complete main audio service (CM) as defined by ATSC Standard A/52 and shall contain the complete primary audio of the program including all required voice-overs and emergency messages.
Data_PID	base_PID+0x000A	PID for the bit stream containing the data for the program.

Table 3-2

Name	PID value
base_PID	0x0340
PMT_PID	0x0340
Video_PID	0x0341
PCR_PID	0x0341
Audio_PID	0x0344
Data_PID	0x034A

Table 3-3

This paradigm enables immediate access to services defined in the paradigm once the program number is known. In order to obtain the PIDs for services not defined by the paradigm but included in the program (for example; additional audio service types or a second data channel), the program_map_table must be decoded. There is a reason that the audio bit stream identified by the program paradigm be a complete main (CM) channel, including all required voice-overs and emergency messages. It ensures that simple receivers, capable of decoding only a single audio bit stream will always receive a complete audio program. Access to all other audio services will require that the program_map_table be decoded to obtain the PIDs for these services.

Examples of services for which a receiver will have to decode the program_map_table include services that can be decoded by simple receivers such as a second CM channel in another language, as well as services that can be decoded by receivers that are able to simultaneously decode and add together two or more audio bit streams, such as music and effects (ME) plus dialogue (D).

According to the program paradigm, every 16th PID is a PMT_PID and may be assigned to a program. If a PMT_MID is assigned to a program by the program paradigm, the next 15 PIDs after that PMT_PID are reserved for elements of that program and are not allowed to be otherwise assigned.

There are some further constraints:

1. Only one program is described in a PSI transport bit stream corresponding to a particular PID value. A transport bit stream containing a program_map_table may not be used to transmit any other kind of PSI table (identified by a different table_id).

2. Adaptation headers are not allowed to occur in transport packets of the PMT_PID for purposes other than for signalling with the discontinuity_indicator that the version_number may be discontinuous.

Repetition Rates

Section zero of the program_association_table may not occur less frequently than every 100 milliseconds.

A program_map_table containing program information may not occur less frequently than every 400 milliseconds.

THE PES PACKET FORMAT

The packetized elementary stream (PES) packet may be generated by either the application encoder or the transport encoder. For the purpose of this explanation, we'll assume the PES encoding to be a function of the transport layer. As was mentioned before, some elementary bit streams, including compressed video and compressed audio go through a process of PES layer packetization, before the transport layer packetization. The PES header includes various rate, and timing information, and information that describes the data, as set by the source encoder. The PES packetization interval will vary depending on the application.

The PES packets that result from this process vary in length. When the PES packet-length field is set to its maximum value, the packet will be 2^{16} bytes (65,536 bytes). In the case of the video stream, the packet length field is set to zero, indicating that the packet length not constrained, and that the header information cannot be used to skip over that packet. The PES packet format will also be used as an input bit stream for digital storage media (DSM).

It should be noted that the format for carrying the PES packet in the transport layer is a subset of the general definition in MPEG-2. This approach was followed to simplify the design and manufacture of the DTV receiver and to make error recovery easier.

A PES Packet

A PES packet includes a PES_packet_start_code, PES header flags, PES header fields, and a payload (data block), as shown in *Figure 3-8*. The payload is created by the application encoder. The payload packet is a stream of contiguous bytes of a single elementary stream. In the case of video and audio packets, the payload is a sequence of access units from the encoder. An access unit is essentially either a video frame or an audio frame.

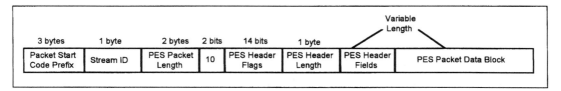

Figure 3-8

A stream_id, which has a unique value, is used to identify each elementary stream. The PES packets that are formed from elementary streams supplied by each encoder carry the stream_id that corresponds to that elementary stream. PES packets that are formed from various types of elementary streams can be multiplexed to form a program or transport stream as described in the MPEG-2 Systems standard.

PES packets for video, including new PTS and DTS values, occur once every picture (or video access unit). The concepts of PTS and DTS have not been defined before, so they will be discussed here. A PTS (*presentation time stamp*) is a field that may be present in a PES packet header that indicates the time at which a presentation unit is presented in the system target decoder. A DTS (*decoding time stamp*) is a field that may be present in a PES packet header that indicates the time that an access unit is decoded in the system target decoder. The PES packets are constructed so that the first bytes of the payload of a video PES packet will be either a sequence, a group of pictures (GOP) or a picture start code. Moreover, a new transport packet will always start with new PES data, and stuffing bytes are used in the adaptation header of the transport packets to make sure that PES packets will always end at the end of a transport packet.

PES Header Flags

The PES header carries information that describes the properties of the bit stream, and that indicate if there are any additional fields in the PES header. *Figure 3-9* shows the flags that are present in the header. Any flags not supported by the digital television system are set to zero.

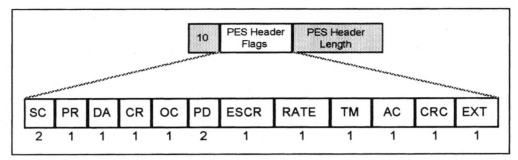

Figure 3-9

The PES Header

The PES header comes directly after the PES_header_length field, which indicates the header size in bytes. The value of the header size includes all of the header fields, and extension fields, and any stuffing_bytes. All of the fields of the PES header are optional, actually. The flags that we just talked about show which fields the header contains, and which fields it doesn't contain. Which fields will be included depends on the particular application. For example, if this is a video packet, the data_alignment_indicator will be set. In this case, the trick mode flag is not set. The application encoder is the entity that provides the information about which flags need to be set, and to encode the corresponding fields.

When the packets being transmitted are PES packets for video elementary streams, the header fields in the PES are organized as shown in *Figure 3-10*. If a long string of zeros were transmitted, which is entirely within the realm of possibility, this could be interpreted as a start code. In order to keep a long sting of 0's from occurring, except when a start code is actually sent, marker bits (more on this later) will have to be inserted.

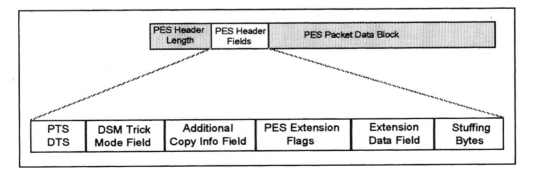

Figure 3-10

PTS AND DTS

We've just alluded to these concepts. PTS is the abbreviation for presentation_time_stamp, which informs the decoder of the intended time of presentation of a presentation unit. DTS stands for decoding_time_stamp, which informs the decoder of the intended time of decoding of an access unit. An access unit is a presentation unit that has been encoded. When it has been encoded, the PTS is associated with the presentation unit that corresponds to the first access unit in the packet. If there is no access unit in a PES packet, the header won't contain a PTS. If the first byte of the synchronization word of an audio frame is present, then the access unit in the packet is an audio access unit. If the access unit in the packet is a video access unit, then it may be a video sequence, a group of pictures (GOP) or a picture header. Under normal conditions, the DTS may be derived from the PTS.

If the value of the DTS is the same as the value of the PTS, the DTS will not be encoded. This situation occurs when the frame is a B-frame. Interestingly, it will also occur when there are no B-frames in the sequence. The reason it occurs when there are no B-frames is in this case there is no reordering delay. The DTS will never occur when by itself, only when there is a PTS present. The opposite situation is not true, however: the PTS may occur when there is no DTS present.

Figure 3-11 is an illustration of the organization of the PTS field when there is no DTS. *Figure 3-12* is an illustration of the orgahization of the PTS field and the DTS field when both are present.

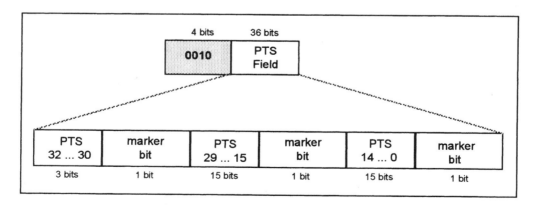

Figure 3-11

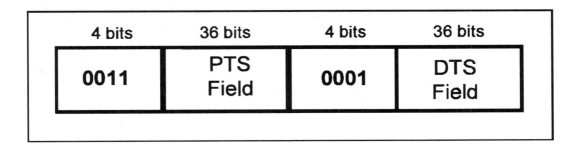

Figure 3-12

PTS Extension Flags

Sometimes further extensions to the PES header exist. In such a case, the EXT flag bit will be set and the header will contain additional flags. If the header field is present, the flag will be set to 1.

CONDITIONAL ACCESS

Sometimes a broadcast station, or a cable company may want to make a program unwatchable to anyone who is not authorized to watch it. The ability to make a program accessible only to certain receiving sets is called "conditional access." The transport protocol in the digital TV system makes conditional access possible. The ability to make access conditional in the system is flexible, and complete in that it supports all transmission aspects of applicable key encryption and scrambling approaches that may be used. Moreover, conditional access is flexible in that it can be applied to each elementary stream, allowing the broadcaster to scramble the video, but not the audio, or vice versa.

With conditional access, the transmitted data is randomized so that an unauthorized decoder can't decode the signal. Receivers that are authorized receive a "key" that is used to initialize a circuit that inverts the bit randomization.

Since this is a somewhat specialized area of the subject of HDTV, this book, at least this edition, won't delve deeply into conditional access.

4
RF/TRANSMISSION SYSTEM

The digital television transmission system has to be designed so that the existing NTSC signal will not interfere with it. It must also be designed so that it will not interfere with the existing NTSC signal. That makes for a very complex method of transmission. The vestigial sideband (VSB) system, which is the transmission system chosen for digital television, can be used in one of two possible modes: a simulcast terrestrial broadcast mode, and a high data rate mode. The term "simulcast," in this case, means that the high-definition digital television signal is being broadcast at the same time as is the NTSC signal that carries the same "program."

In other words, in those areas in which a broadcast station is currently broadcasting the HDTV signal for, say, *The Tonight Show*, for people who have HDTV receivers, the same station is, at exactly the same time (simultaneously), broadcasting the same program in the NTSC mode for those of us with more modest pocketbooks, and our tried and true NTSC receivers.

The high data rate mode transmission would be for transmission systems such as cable TV.

The terrestrial broadcast mode and the high data rate mode share the same pilot, symbol rate, data frame structure, interleaving, Reed-Solomon coding, and synchronization pulses. The terrestrial broadcast mode provides a single ATV signal in a 6 MHz channel, and is set up so that it will reach as broad a geographic area asp possible. The high data rate mode, on the other hand, which trades off some robustness for twice the data rate, can carry two ATV signals in a single 6 MHz channel.

Because it is important in both modes that the receiver be able to use the signal to produce useful audio and video for display on the screen and via the loudspeakers, both modes utilize a pilot, a segment sync, and a "training sequence" for robust acquisition and operation. Both system modes also use exactly the same carrier, sync and recovery circuits, as well as phase correctors and equalizers. Moreover, both modes use the same Reed-Solomon (RS) code for forward error correction (FEC).

In order to get a usable HDTV signal to as broad an area as possible, the broadcast mode uses a process called "trellis coding", and the HDTV receiver circuitry includes an NTSC rejection filter that utilizes this trellis coding. The trellis code incorporates precoding at the transmitter. When the NTSC rejection in the receiver is activated, the trellis decoder is switched to a trellis code that corresponds to the encoder trellis code that is concatenated with the filter.

The high data rate mode, on the other hand, is transmitted over a channel that is more or less protected. It doesn't have to contend with NTSC interference or other interference that takes place in the atmosphere, so the higher data rate is transmitted in the form of more data levels (bits per "symbol"). And when a receiver receives one of these signals, there is no trellis coding, and the NTSC interference rejection filter is turned off.

Because of the nature of VSB transmission, the receiver only needs to process the inphase (I) channel signal, sampled at the symbol rate, thus making it less costly to design and build the receiver. The decoder requires only one A/D converter, and a real (as opposed to complex) equalizer operating at the symbol rate of 10.76M sample/s. The parameters for the two VSB transmission modes are shown in *Table 4-1*.

Parameter	Terrestrial mode	High data rate mode
Channel bandwidth	6 MHz	6 MHz
Excess bandwidth	11.5%	11.5%
Symbol rate	10.76 Msymbols/s	10.76 Msymbols/s
Bits per symbol	3	4
Trellis FEC	2/3 rate	None
Reed-Solomon FEC	T=10 (207,187)	T=10 (207,187)
Segment length	832 symbols	832 symbols
Segment sync	4 symbols per segment	4 symbols per segment
Frame sync	1 per 313 segments	1 per 313 segments
Payload data rate	19.28 Mbps	38.57 Mbps
NTSC co-channel rejection	NTSC rejection filter in receiver	N/A
Pilot power contribution	0.3 dB	0.3 dB
C/N threshold	14.9 dB	28.3 dB

Table 4-1

Notice that the table includes the term "symbol." This is a quite nebulous term. Remember, we're talking about digital modulation, here. We're modulating a radio signal using a digital signal, rather than an analog signal. For purposes of this discussion, it's useful to think of a "symbol" as a collection of data that is transmitted by an ATV transmitter, and received and processed by an HDTV receiver, and which represents a unit for the purposes of transmission and reception, but not necessarily any kind of entity as regards the display of video or audio.

The Receiver

This is the end of the HDTV system that most of us are interested in. Delivery of HDTV receivers is still in the early stages, and so, solid technical information and schematic diagrams are very scarce. So, we'll present a conceptual receiver based on the system established by the so-called "Grand Alliance." *Figure 4-1* shows the receiver block diagram of the VSB terrestrial broadcast transmission system.

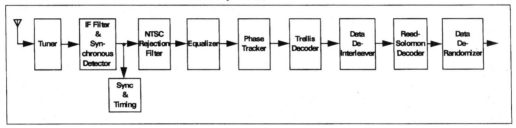

Figure 4-1

The Tuner

The tuner for this receiver, which was the prototype submitted for testing the HDTV system, is shown in *Figure 4-2*. The tuner receives the UHF or VHF 6 MHz signal from the antenna. It is a high-side injection, double conversion type with a first IF frequency of 920 MHz. This type of design places the image frequencies above 1 GHz, making it easy to reject them using a fixed front-end filter.

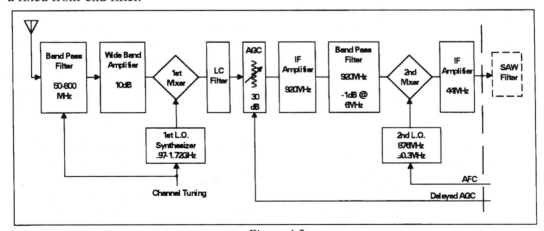

Figure 4-2

By selecting 920 MHz as an IF frequency, the designers reaped a two-fold benefit:

- It is high enough so that the selevity of the input bandpass filter prevents the output of the local oscillator (978 MHz to 1723 MHz) from leaking out the tuner front end and interfering with other UHF channels.

- It is low enough so that second harmonics of UHF channels (470 MHz to 806 MHz) are above the first IF bandpass.

It is possible that harmonics of cable channels could occur in the first IF passband, but that is not seen to be a significant problem because cable systems transmit a signal that is relatively flat (+/-10 dB) across the frequency spectrum, and is at a relatively low signal level (-28 dBm or less).

A bandpass filter at the tuner input limits the frequency range to 50 MHz to 810 MHz. This causes the tuner to reject any nontelevision signals that might be within the tuner's image frequency range (above 920 MHz). Other television signals are rejected by a broadband tracking filter. This tracking filter does not have a narrow bandwidth, and it is not critically tuned. This is unlike the case with present day NTSC tuners that must reject image signals only 90 MHz away from the desired channel. This filter introduces little or no channel tilt.

From the bandpass filter, the received signal goes to a wideband RF amplifier with a gain of 10 dB. This amplifier increases the level of the signal as it is fed into the first mixer. It is the most significant factor affecting receiver noise figure, which is 7 dB to 9 dB over the entire VHF, UHF and cable bands.

The first mixer is a highly linear double-balanced unit designed to minimize generation of even harmonics, is driven by a synthesized low phase noise local oscillator (LO) whose output frequency is above the first IF frequency (high-side injection). A microprocessor is used to control both the channel tuning (first LO) and broadband tracking filtering (input bandpass filter). The tuner is capable of tuning the entire VHF and UHF broadcast bands as well as all standard IRC and HRC cable bands.

From the mixer, the signal goes to an LC filter in tandem with a narrow 920 MHz bandpass ceramic resonator filter. The LC filter selects against the spurious harmonic and subharmonic responses of the ceramic resonators. The 920 MHz ceramic resonator bandpass filter has a bandwidth of about 6 MHz, and causes -1 dB of attenuation. A 920 MHz IF amplifier is placed between the two filters. An AGC circuit with a 20 dB range protects the active stages that follow from being overloaded by a large signal.

The output of the bandpass filter is then fed to a second mixer, which is driven by a second LO, an 876 MHz voltage-controlled SAW (surface acoustic wave) oscillator. This LO is controlled by the frequency-and-phase-locked loop synchronous detector. The output of the second mixer is 44 MHz, which is the desired second IF frequency. This signal is fed to a constant gain 44 MHz amplifier.

The output of the tuner feeds the IF SAW filter and synchronous detection circuitry.

The tuner is fabricated from standard consumer electronics components and is housed in a stamped metal enclosure.

Channel Filtering and Recovery of the VSB Carrier

Since the carrier frequency is not transmitted in vestigial sideband transmission, but it is needed to extract the modulating signal from the received signal, it must be recovered at the receiver. The means to accomplish this carrier recovery is transmitted along with the HDTV signal, a small pilot carrier. The FPLL (frequency and phase-locked loop) circuit of *Figure 4-3* is used to perform this carrier recovery.

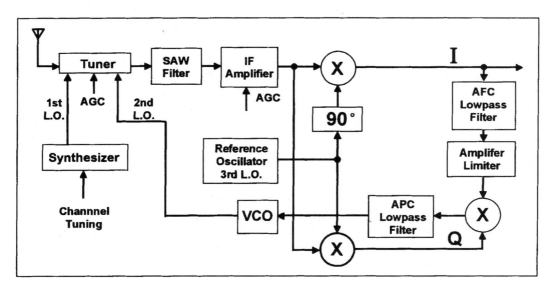

Figure 4-3

The first LO is synthesized by a PLL (phase-locked loop) and controlled by a microprocessor. The third LO is a fixed-reference oscillator. Any frequency drift or deviation from the normal is compensated in the second LO. Control for the second LO comes from the FPLL synchronous detector. This detector contains both a frequency loop, and a phase-locked loop in a single circuit. The frequency loop can lock to any frequency within +/- 100 kHz of its nominal frequency. The phase-locked loop has a narrow bandwidth: less than 2 kHz.

When the set is in the process of tuning one received signal from all of the broadcast or cable channels available (frequency acquisition) the frequency loop uses both the in-phase (I) and quadrature-phase (Q) pilot signals. All of the other data processing circuits in the receiver use only the I channel signal.

The set has not achieved phase lock on the desired channel:

- When the set is first turned on.

- After the signal has been lost.

- After the viewer has changed the channel.

During this time, the AFC (automatic frequency control) low-pass filter acts on the beat signal created by the mixing of the incoming pilot signal with the VCO signal. The high-frequency signal is mostly rejected by the AFC filter, leaving only the pilot beat frequency.

This pilot beat signal is limited to a constant amplitude square wave (+/-1) and then used to multiply the quadrature signal. The result is a bipolar S-curve AFC characteristic. The polarity of the S-curve error will depend on whether the VCO frequency is above or below the frequency of the incoming IF signal. This DC signal is filtered and integrated by the automatic phase control low pass filter, and is then used to adjust the tuner's second LO to reduce the difference between the output frequency of the second LO and the incoming signal.

When the difference in frequency is close to zero, the APC loop takes over and phase-locks the incoming IF signal to the frequency of the third LO. This is a normal phase-locked loop circuit, with the added feature that it is biphase stable. In this circuit, the correct phase-lock polarity is determined by forcing the polarity of the pilot to be the same as the known transmitted positive polarity.

Once locked, the detected pilot signal is constant, the limiter that feeds the third multiplier is at a constant +1, and the frequency loop is automatically disabled. Thus, only the phase-locked loop is in operation. The bandwidth of the APC low-pass filter is broad enough to allow frequency pull-in reliably throughout +/-100 kHz. At the same time, the bandwidth of

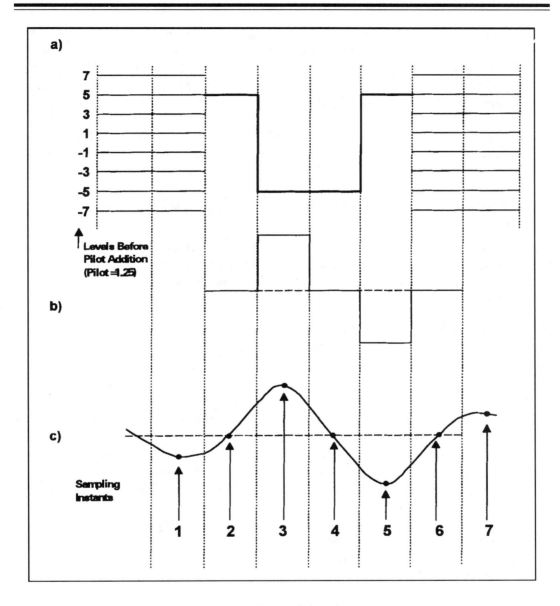

Figure 4-4

this filter is narrow enough so that it consistently rejects all strong white noise and NTSC co-channel interference signals. Interestingly, the white noise rejected by this filter includes data signals.

The bandwidth of the PLL is narrow enough so that it will reject most of the AM and PM generated by the data. At the same time, the PLL's bandwidth is sufficiently broad that it can track out any phase noise on the signal (and therefore on the pilot) to about 2 kHz. Tracking out low frequency phase noise (as well as low frequency FM components) improves the effectiveness of the phase tracking loop.

The prototype receiver that was used in demonstrating the HDTV system was able to acquire a signal and maintain lock at a signal-to-noise ratio of 0 dB or less, in the presence of a great deal of interference.

GETTING THE PROGRAM DATA OUT OF THE SIGNAL

The data sync information (*Figure 4-4*), which is repetitive at specific intervals, is selected from among the synchronously detected random data by a filter that has a narrow bandwidth. From these data syncs, a properly phased 10.76 MHz symbol clock is created as well as a coherent AGC control signal. The circuit that performs these functions is shown in *Figure 4-5*.

After it is detected by the synchronous detector, the I channel composite baseband signal, which contains the data bit streams for the program that was transmitted, is converted by an A/D converter for digital processing. A PLL is used to derive a 10.76 MHz symbol clock for the receiver.

The circuits that follow this stage operate on the signal to make sure that they have the proper synchronization so that the transmitted data can be successfully and properly extracted from the transmitted signal. Once it has been established with the required level of confidence that the synchronization is such that the remainder of the receiver circuits can use the extracted data to create and display the appropriate audio and video, these circuits are enabled.

Data segment sync detection and clock recovery both work reliably at signal-to-noise ratios of 0 dB or less, even if there is a great deal of interference.

Noncoherent and Coherent AGC

Quite a bit of control of the incoming signals is provided by automatic gain control (AGC) circuits in the receiver. Before carrier and clock synchronization are effected, *noncoherent* AGC is performed whenever any signal is too large for the A/D converter to handle. The signals whose amplitudes are thus controlled include any locked or unlocked signal, or noise or interference. To achieve this gain control, the IF and RF gains are reduced, with the appropriate AGC "delay" applied.

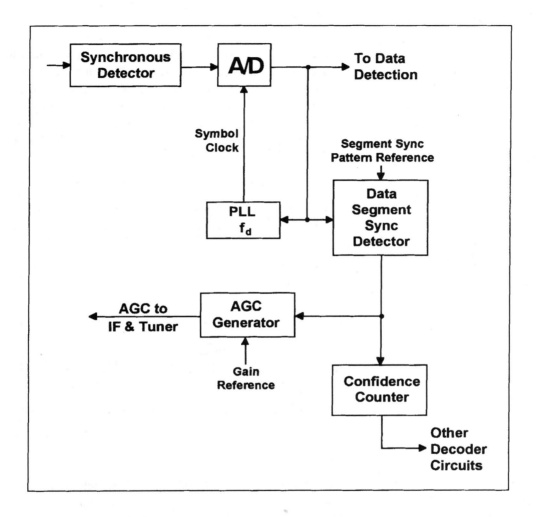

Figure 4-5

Once data segment syncs have been detected, *coherent* AGC is applied, using the amplitudes of the segment syncs as reference. Going back to the transmitter, at that end the amplitude of the bipolar syncs relative to the amplitude of the discrete levels of the data to be transmitted is determined. After the syncs are detected by the circuitry of the receiver, their amplitudes are compared to the amplitude of a reference value. The difference between the two amplitudes, the error, is integrated. The output of the integrator controls the IF and "delayed" RF gains, establishing them at whatever level yields the correct sync amplitudes.

Synchronization of the Data Fields

As the received data exits the A/D converter, it is filtered by an interference rejection filter to minimize co-channel interference. From there, each received data segment is compared to ideal field No. 1 and ideal field No. 2 reference signals (*Figure 4-6*). This achieves Data Field Sync detection. Because a precision data segment and symbol clock have already been reliably created by the clock recovery circuit, oversampling of the field sync is not necessary. As a result, the field sync recovery circuit knows exactly where a valid field sync correlation should occur within each data segment, so it only needs to perform a symbol by symbol difference.

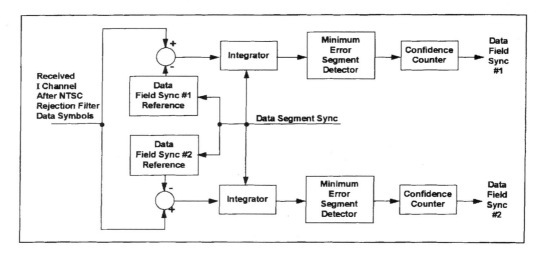

Figure 4-6

Once the receiver circuits have reached a preset level of confidence that field syncs have been detected on given data segments, the Data Field Sync signal is made available to following circuits in the receiving set. Field sync recovery is reliable, even under adverse signal to noise ratios, and serious interference conditions.

Interference Rejection Filter

Digital television receivers use a comb filter to reject interference from NTSC signals. Let's start by having a look at the NTSC signal. *Figure 4-7* shows the three principal components of the NTSC signal: the visual carrier (V), which is located 1.25 MHz from the lower edge of the band, the chrominance subcarrier (C) which is located 3.58 MHz higher than the visual carrier frequency, and (3) the audio carrier (A) which is located 4.5 MHz higher than the video carrier frequency.

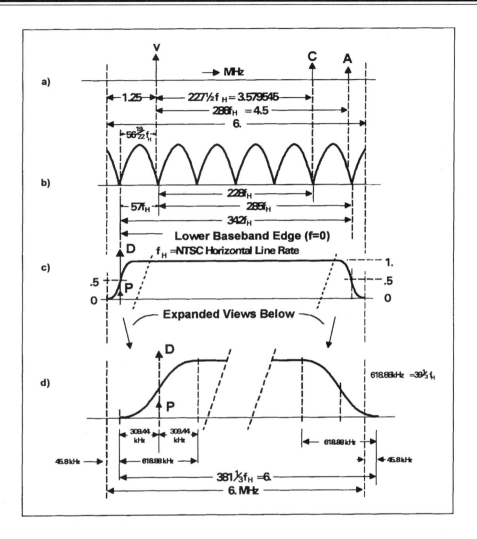

Figure 4-7

The NTSC interference rejection comb filter (*Figure 4-8*) is a single tap, linear feed-forward filter. The frequency response of this filter is illustrated in *Figure 4-7*. A characteristic of this filter is the presence of nulls spaced $57 * f_H$ (10.762 MHz/12, or 896.85 kHz) apart. There are seven nulls within the 6 MHz channel. The frequency of the NTSC video carrier is close to the frequency at the second null from the lower band edge of the comb filter. The sixth null of the comb filter is designed to occur precisely at the frequency of the chrominance subcarrier of the NTSC signal. The seventh null from the lower edge of the band occurs near the frequency of the NTSC audio carrier.

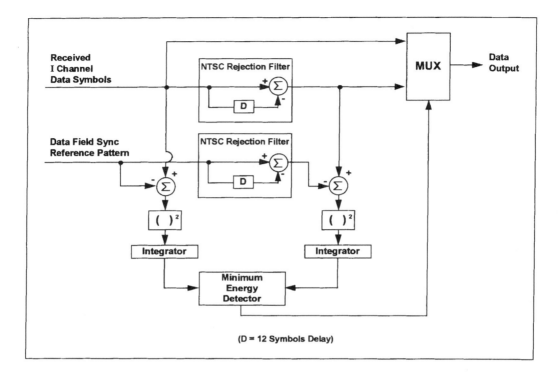

Figure 4-8

If you compare parts a and b of *Figure 4-7*, you will see that:

• The video carrier frequency is 2.1 kHz below the second null of the comb filter.

• The frequency of the chroma subcarrier falls near the sixth null.

• The audio carrier falls 13.6 kHz above the seventh null.

The amplitude of the NTSC audio carrier is at least 7 dB below that of its video carrier.

Unfortunately, the comb filter does not filter out all possible NTSC interference. It does filter out any steady-state signals located at its null frequency points, but it has a finite response time. Therefore, if the NTSC signal that could interfere with the program being tuned by the receiver has a sudden abrupt change in carrier level (as at the leading and trailing edges of the sync pulse) some interference will be admitted to the set. This interference will be one cycle of the zero-beat frequency (offset) between the ATV and NTSC carrier frequencies. The amplitude of this interfering signal will be proportional to the NTSC step size.

If the amplitude of this interference is large enough, some errors, called "data slicing" errors, will occur. Error correction circuits in the receiver should be able to correct for these problems, however.

The interference filter is configured so that it will be on during times when it is needed to reject interference from NTSC signals, but switched off when no such interference is present. The circuit of *Figure 4-9* can detect NTSC interference. The signal-to-interference-plus-noise ratio of the binary Data Field Sync is measured at the output of the comb filter, and compared to the value that occurs at the input of the filter. The path that has the largest signal-to-noise ratio, either the path around the filter or the path through the filter, is switched in automatically.

It would require less circuitry and complication to keep the NTSC rejection comb filter switched in all the time. However, the filter introduces its own degree of degradation of the signals, and so it's better for it to be out of the circuit when it's not needed. While it rejects

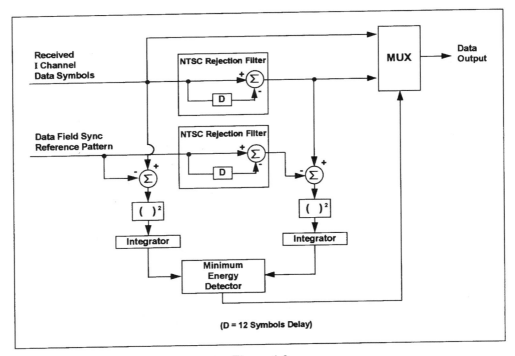

Figure 4-9

NTSC co-channel interference, the rejection comb filter degrades white noise performance by 3 dB, because the filter output is the subtraction of two full-gain paths. The result is that because white noise is not correlated from symbol to symbol, the noise power increases by a factor of two. An additional 0.3 dB degradation is introduced because of 12 symbol differential decoding by the trellis decoder.

To make the signal as strong and usable as possible at all times, the NTSC interference rejection comb filter is switched out when it is not needed. At some time in the future, as we know, broadcasts using the NTSC system will be eliminated entirely. When that happens, HDTV set manufacturers will no longer have to include these filters in their sets.

Channel Equalizer

Contrary to popular opinion, problems such as ghosts can, in fact, affect digital television, so the circuitry of an HDTV receiver includes an equalizer/ghost canceller to compensate for linear channel distortions such as tilt and ghosts. These distortions can be the result of problems in the received signal, or they may arise because of less than perfect components in the set.

We won't go into the operation of the equalizer in detail here, but its block diagram is shown in *Figure 4-10*.

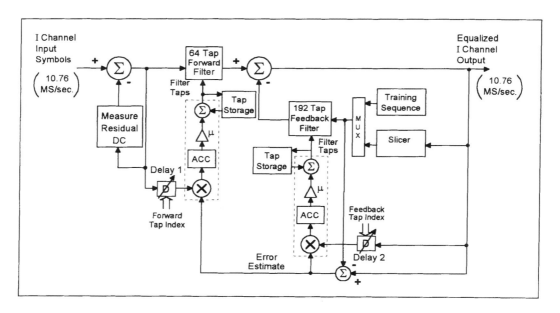

Figure 4-10

Phase Tracking Loop

The phase tracking loop further tracks out phase noise that has not been eliminated by the IF PLL operating on the pilot. A block diagram of this circuit is shown in *Figure 4-11*. The gain of the output of the real equalizer operating on the I signal is controlled by a multiplier. This signal is then sent to a filter which creates an approximation of the Q signal. The I and Q signals are fed into a complex multiplier, called a derotator, which removes the phase noise.

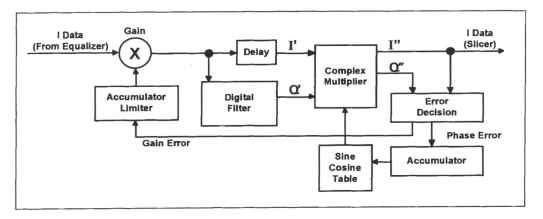

Figure 4-11

Trellis Decoder

At the outset of this discussion, let's make the statement that the trellis decoder is another artifact of the NTSC-compatibility situation. When there are no more NTSC signals for HDTV sets to have to reject, the trellis decoder will be eliminated from receiver circuitry, along with the NTSC rejection filter. However, for the time being, trellis decoders will be part of the circuitry of HDTV sets, and anyone who wants to be familiar with these sets will have to be familiar with them.

The concept of trellis decoding is abstruse in the extreme, a matter for mathematicians and data transmission experts, so it won't be treated in depth here, but since trellis decoding takes place in today's HDTV receivers, it has to be at least mentioned. The concept of the trellis code deinterleaver is shown in *Figure 4-12*.

Before the 8 VSB signal can be processed by the trellis decoder, the Segment Sync has to be suspended. The circuit shown in block diagram form in *Figure 4-13* effects suspension of Segment Sync.

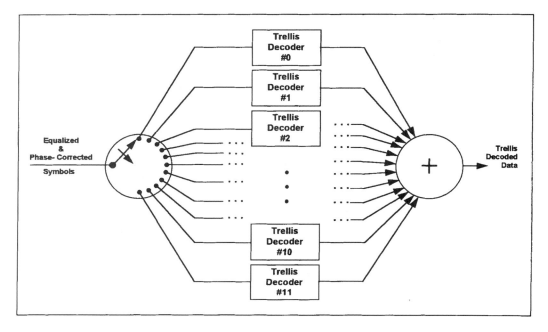

Figure 4-12

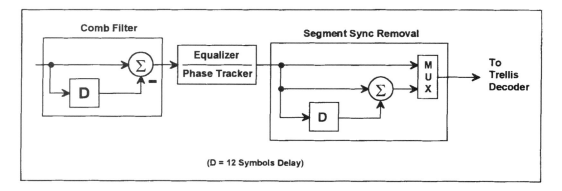

Figure 4-13

At the transmitting end, the data is encoded in such a way that errors that may have been caused as the signal traverses the transmission path can be detected at the receiver. What happens in the trellis decoder is that the decoder generates all possible code sequences that are to be compared with the received sequence. It eliminates the unlikely sequences, and by so doing, it tries to keep track of the encoder state as a function of time. A "trellis" is a way of visualizing a path through all of the encoder states over time. The term "trellis" is used because the path through the possible states resembles in a vague way a garden trellis used to train the growth of climbing flowers.

The trellis decoder performs the task of slicing and convolutional decoding. It has two modes. One mode is employed when the NTSC rejection filter is used to minimize NTSC co-channel interference, and another mode that is used when the NTSC rejection filter is turned off (*Figure 4-14*).

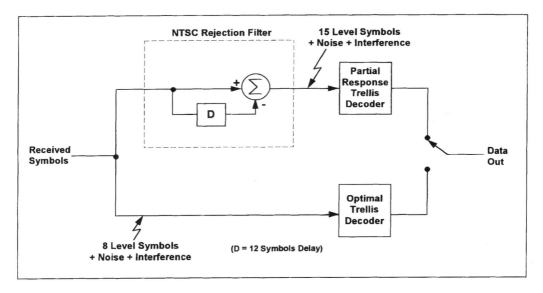

Figure 4-14

Data Deinterleaver

At the transmitting end, the transmitter uses a convolutional interleaver to interleave data to improve its susceptibility to noise. At the receiver, the received data has to be deinterleaved. This function is performed in a convolutional deinterleaver, shown in conceptual form in *Figure 4-15*.

Reed-Solomon Decoder

Once the data has been trellis decoded and deinterleaved, it is sent to a circuit called a Reed-Solomon decoder which uses bytes of information called "parity bytes" that were transmitted along with the program data to correct for errors that were introduced by the transmission

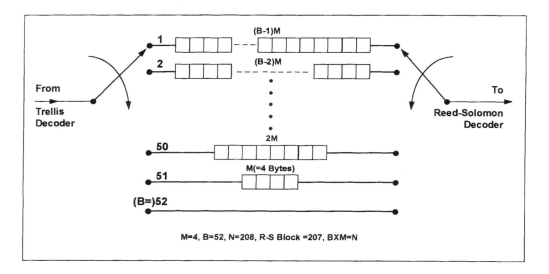

Figure 4-15

path. Many types of noise, such as impulse noise, NTSC co-channel interference, and trellis-decoding errors will be reduced considerably by the use of interleaving and Reed-Solomon error correction.

Data Derandomizer

For purposes of transmitting the best possible signal, the data is randomized at the transmitter by a process called a pseudo random sequence, and there has to be an equal and opposite derandomizing process at the receiver.

Receiver Loop Acquisition Sequencing

Whenever the viewer turns on the set, or changes channels, the set goes through a "universal reset," in which the various loops in the set acquire and lock up in sequence. In this sequence, each loop is independent of loops that have already occurred:

• Acquisition of the 1st local oscillator synthesizer

• Noncoherent AGC reduces unlocked signal to within the input range of the A/D converter

• Acquisition of the carrier (FPLL)

- Data segment and clock acquisition

- Coherent AGC of signal (IF and RF gains set properly)

- Data field sync acquisition

- Decision as to whether to connect the NTSC rejection filter

- Equalizer completes tap adjustment algorithm

- Trellis and Reed-Solomon data decoding begin

Most of these loops mentioned have "confidence counters" associated with them to assure proper operation.

THE REST OF THE STORY

The description of the HDTV receiver up to this point concerns what is called the "front end," the portion of the unit that tunes in the signal and decodes it, pulling the data that represents the pictures and sound out of the air, and turning it back into video and audio. Once that has been accomplished, the remainder of the circuitry in the set is very similar to the circuitry in the now familiar NTSC sets.

High-definition sets have all the normal circuits that it takes to make a TV set work: power supply, horizontal deflection, vertical deflection, synchronization, audio amplifier, speakers, and, of course, that 16 x 9 screen. Very little information about the specifics of these circuits is available as this book goes to press, but it should be available soon, and will be included in the next edition.

5

HISTORY OF CONSUMER ELECTRONICS LEADING UP TO HDTV

In order to gain a little perspective on HDTV, it's useful to look at some of the events that have taken place recently that brought HDTV to fruition, as well as to look at some of the milestones that have taken place in the field of consumer electronics that brought us to the threshold of the DTV era. To that end, we present some of the documentation from the FCC, the Electronic Industries Association, and others that have led us to HDTV.

This is an unofficial announcement of Commission action. Release of the full text of a Commission order constitutes official action. See MCI v. FCC. 515 F 2d 385 (D.C. Circ 1974).

Engineering & Technology Action

Report No. ET 98-10

November 16, 1998

FCC RELEASES DIGITAL TELEVISION CONSUMER BULLETIN

The FCC's Office of Engineering and Technology today issued a Digital Television (DTV) Consumer Information Bulletin to provide consumers with more information on what promises to be one of the most significant developments in television technology since the advent of color television in the 1950s.

The Consumer Bulletin will provide consumers with information on the new DTV technology, its upcoming deployment, developments such as improvements in the compatibility between digital programming and the equipment used by broadcasters and cable operators, and the capabilities and features that are expected to be available in the new DTV sets.

This bulletin, and updated information on DTV, is available on the FCC Web site at http://www.fcc.gov/oet.

November 1998

The arrival of digital television ("DTV") this fall promises to be one of the most significant developments in television technology since the advent of color television in the 1950s. DTV has the capability to provide clearer and sharper, cinema-like pictures as well as multi-channel, CD-quality sound. It can provide new uses such as multiple video programs or other services on a single television channel, including data services. The use of DTV technology will also allow television to enter the digital world of the personal computer and the Internet.

As with any major technology change, it will be important for consumers to understand the capabilities of new equipment in order to make purchase decisions. The new digital television sets will have many new features and technical characteristics that will vary somewhat between different models and manufacturers. In addition, special features may be needed when using DTV sets to receive programs from cable, direct satellite, or other video service providers.

This bulletin has been prepared to provide consumers with information on the new DTV technology, its upcoming deployment, and the capabilities and features that are expected to be available in the new DTV sets. We will release periodic updates to this bulletin to help consumers keep up with digital television developments such as improvements in the compatibility between the digital programming and equipment used by broadcasters and cable operators.

1. Why change today's TV system to DTV?

Today's television system has served the American public well for almost three generations. Over the years, additional features and functions, such as color, stereo sound, remote control, cable channels, closed-captioning, and parental control features were developed. Even with these improvements, TV sets became less expensive and more affordable.

However, today's television system is now showing its age. A transition to digital is occurring in all media technologies. Direct broadcast satellite providers, for example, have offered digital service since their inception. In contrast, the shortcomings of the 50-year-old analog technology used by broadcasters—such as limited resolution and color rendition, as well as problems with "ghosts" and interference from other radio sources—have become increasingly apparent as consumer TV sets became larger and more technologically advanced.

Therefore, at the request of the television broadcast industry, the FCC sponsored an effort to investigate the possibility of significantly improving the technical quality of today's television system.

More than 10 years ago, scientists and engineers in the broadcast, cable, and consumer electronics industries and in government worked on the development of this new television system. Many different approaches were developed and tested. Today's DTV system represents the culmination of this work. DTV eventually will replace today's analog television service. After a transition period that allows stations to construct DTV transmission facilities and consumers gradually to replace their TV sets, broadcasts using the existing analog television system will cease and all over-the-air broadcast television service will be provided with the new DTV system.

2. What is DTV?

DTV is a new "over-the-air" digital television system that will be used by the nearly 1,600 local broadcast television stations in the United States. The DTV standard is based on the Advanced Television System Committee (ATSC) standard A/53. (For more details on this standard, you can also visit the ATSC Web site at http://www.atsc.org/.)

The DTV standard is a very flexible television system that will allow broadcasters to provide new and higher quality services. First, DTV will permit transmission of television programming in new wide-screen, high-resolution formats known as high-definition television (HDTV). In addition, the new DTV television system allows transmissions in standard-definition television (SDTV) formats that provide picture resolution similar to existing television service. Both the HDTV and SDTV formats will have significantly better color rendition than the existing analog television system. The DTV system also will allow broadcasters to transmit multiple programs simultaneously using a single television channel. TV stations will, depending on the type and source of programming, be able to transmit multiple SDTV programs or in some cases two HDTV programs. DTV also will provide improved audio quality, similar to that of compact discs, with up to five channels of sound per program.

The new system will also support delivery of digital data services simultaneously with television and audio programming. Using this data transmission capability, it will be possible for broadcast stations to send publications (such as a local "electronic newspaper"), program schedules, computer software, information requested about specific products, or virtually any other type of information, at the same time that they transmit regular television programming. The DTV system also provides the flexibility to support the introduction of new services in the future, as technology and viewer interests continue to develop.

3. How does DTV provide these improvements?

With DTV, television pictures, sound, and new data services will be transmitted digitally, rather than as an analog signal. The increased capabilities and new services of DTV are made possible through the use of digital compression techniques that allow more information to be transmitted in the same amount of spectrum used by an existing television channel. The data rate of the DTV signal in the 6 MHz broadcast television channel is 19.44 Mbps. This compares with data rates of today's telephone modems of between approximately 28 and 56 Kbps.

4. How will DTV be provided by broadcasters during the transition?

Television stations will operate two channels during the transition: an existing analog channel as well as a new DTV channel. The analog channel will allow consumers to continue to use their current TV sets to receive traditional analog programming during the transition. The DTV channel will allow consumers to receive new and improved services with new DTV sets or with special converter boxes that will allow some DTV programs and services to be viewed on existing analog sets. Roughly half of the nation's households should be able to receive DTV service by the end of 1999, and everyone else will have access by 2002 (more information on this schedule is provided below). At the end of the transition—which is currently scheduled for 2006—stations will relinquish one of their two channels as they cease analog service and make the change to all-DTV service. The transition schedule is subject to periodic review by the FCC and can be extended beyond 2006 if certain conditions relating to the number of consumers who can receive DTV signals are not met.

5. When will broadcasters begin providing DTV service?

Local broadcasters will be initiating DTV service at different times. A station may begin DTV service as soon as it has received its FCC permit and is ready with equipment and other necessary preparations. The FCC has established a schedule by which broadcasters must begin DTV service (absent extenuating circumstances that may affect individual stations). This schedule requires that stations affiliated with the top four networks (ABC, CBS, FOX, and NBC) in the 10 largest markets begin service by May 1, 1999. Stations affiliated with these networks in markets 11-30 must begin service by November 1, 1999. All commercial stations must begin DTV service by May 1, 2002, and all noncommercial educational stations must start by May 1, 2003. A number of stations started DTV service in November 1998. These stations are indicated in the chart on the next page:

Stations Providing DTV Service—November 1998*

Market	ABC	CBS	FOX	NBC	Other
New York		WCBS			
Los Angeles	KABC	KCBS		KNBC	KTLA (WBN)
Chicago					
Philadelphia	WPVI	KYW	WXTF	WCAU	
San Francisco	KGO	KPIX	KTVU	KRON	KBHK (UPN)
Boston	WCVB				
(Manchester, N.H.)	WMUR				
Washington	WJLA	WUSA		WRC	WETA (PBS)
Dallas	WFAA		KDFW	KXAS	
Detroit	WXYZ		WJBK		
Atlanta	WSB			WXIA	
Houston			KHOU		
Seattle	KOMO			KING	KCTS (PBS)
Portland, Ore.					KOPB (PBS)
Indianapolis				WTHR	
Charlotte		WBTV			
Raleigh		WRAL			
Cincinnati				WLWT	
Kansas City					KCPT (PBS)
Milwaukee					WMVS (PBS)
Columbus, Ohio	WBNS				
Harrisburg, PA					WITF (PBS)
Honolulu	KITV				
Madison, Wis.	WKOW				
Jackson, Miss.					WMPN (PBS)

* The information in this table is provided courtesy of the National Association of Broadcasters.

6. Will DTV operate for the same hours and reach as far as existing local TV service?

TV stations generally will be required to operate their DTV service on the same schedule as their existing service. Most stations' DTV service will reach at least as far as their existing service. However, some stations may operate initially with lower power transmitters, which may cause DTV service in these individual cases to not reach as far as existing TV coverage. You can contact your local TV station to see if its DTV service will be available where you live. Specifically, you may want to ask your local broadcaster whether its DTV signal is strong enough to cover its entire analog service area. If it is not, you may want to ask your broadcaster whether your neighborhood is within the station's initial DTV coverage area and when the broadcaster plans to extend its DTV signal coverage.

7. Do I need a new TV set to receive DTV?

In general, to enjoy the full benefits of DTV such as wide-screen, higher-resolution pictures you will need to purchase a new DTV set. Existing television sets will not be able to display DTV signals. However, it is expected that less expensive converter boxes will be available that will allow you to watch standard-definition DTV on an existing TV set. These boxes will receive DTV signals and convert them to the transmission system used by existing TV sets. The pictures received through these converter boxes should be clear of the "ghosts" and other interference that are characteristic of today's analog TV service in some areas. These converter boxes also will allow any new DTV programs (i.e., programs that are not also available via traditional analog service during the transition) to be displayed on existing TV sets. However, because most existing TV sets were not designed to display high-resolution pictures, converter boxes will not be able to provide the higher HDTV picture quality that will be available on new DTV sets. Also, it is possible that some new DTV sets may be marketed that will not be able to display all DTV formats. For example, some DTV sets may not be able to display HDTV signals at their full resolution potential or some sets may not have the new wider screen size. Consumers should be aware of these format differences in selecting DTV sets and should ask electronics retailers to fully explain the capabilities of new DTV equipment.

8. Will the new DTV sets be expensive?

Just as color sets were expensive when they were first introduced in the mid-1950s, the new DTV sets will be expensive at first. However, manufacturers have indicated that they expect prices to fall over time. As noted above, consumers also will have the option of obtaining a converter box that adapts an existing set to digital service. The price of these boxes is expected to drop significantly during the transition to full DTV service.

9. Do I need an outside antenna to receive DTV? Is the antenna I use for existing TV reception good enough?

DTV is intended to work with an outside TV antenna (as is today's analog TV service). If you have an outside antenna and it provides acceptable TV reception now on UHF channels (i.e., channels 14-69), it should also work for DTV. Also, if your indoor antenna is capable of receiving UHF television service now, you may also be able to receive DTV service with that antenna. Indoor DTV reception is affected by a number of factors that vary depending on local conditions. Many retailers carrying DTV equipment have information about local reception conditions. In addition, some retailers are providing technical assistance to customers currently purchasing DTV equipment, including in some cases visits to the home to assist with reception issues. You may want to advise your retailer whether you plan to use an indoor or outdoor antenna and ask whether technical assistance is available.

10. Will I be able to receive existing TV programming on my DTV set?

Yes. Digital television sets available during the transition will be fully compatible with traditional analog TV programming. These new digital sets will have the capability to receive new DTV programming and will also be able to receive all the programming you receive today on your traditional set. This means that new DTV sets will be able to display all of the programming available today from broadcasters, cable operators, satellite TV services, other video service providers, and prerecorded sources. So, if you buy a new digital set, you will not lose the ability to get any of the programs you now receive on your current set. You should know, however, that depending on your cable operator's plans for carrying DTV signals (see below), you may need an antenna to receive broadcast DTV programming. To watch both antenna-received programming as well as programming from other sources such as cable or satellite, you may need extra equipment (such as an input selector or "A/B" switch).

11. Will cable systems carry local DTV signals?

The FCC does not currently require cable systems to carry DTV signals. The FCC is, however, considering whether to impose such requirements. Nevertheless, some cable operators have indicated that they may carry the DTV programming of local broadcast stations. If these signals are carried in their original DTV format, no additional equipment will be needed to receive them on DTV sets. Some cable systems may, however, convert DTV programming to different digital formats and/or may carry DTV signals at lower resolutions than the original broadcast signal. In such cases, special cable "set-top" boxes may be needed to receive DTV. Consumers should consult with their local cable operator about its plans for providing the DTV programming of local television stations. Because cable operator plans vary from market to market, consumers seeking additional information may want to pose the following questions to local cable providers:

• Will your cable system carry local digital broadcast channels when they become available in this market?

• If I receive my local digital broadcast programming over cable, will these channels be in the same resolution as if I was receiving them with an antenna?

• Will I need a new cable box to receive local digital broadcast channels?

• If I will need a new cable box, how much will it cost?

• Can I buy the new cable box from a retail outlet, or must it be provided by the cable company?

• If I need a new cable box, how will it affect the features of my new DTV set? For example, will I be able to use picture-in-picture features? Can I use the on-screen program guide that is included in my new DTV set?

12. Some cable systems are offering digital services. Does this mean that these cable systems will provide DTV signals just like broadcasters?

Not necessarily. Many cable systems are now offering digital service. However, these digital systems generally are intended to allow cable systems to provide more channels and other services, such as access to the Internet. Special digital set-top boxes are needed to convert digital cable signals for display on existing TVs. Also, cable systems use different digital standards than the DTV standard used by broadcasters and current DTV sets, so digital cable services currently cannot be directly received by DTV sets. Therefore, just as is the case with many existing TVs, a special digital set-top box will be needed to receive digital cable services on a new DTV set. Consumers should be aware that there may be additional charges for digital cable services and that special set-top boxes are needed to receive them. The FCC has instructed the cable industry and consumer electronics manufacturers to work together to try to develop standards to facilitate compatibility between DTV sets and digital cable service. This eventually may permit DTV sets to receive cable service (including digital service) without the need for special converter boxes.

13. Will there be any HDTV cable programming available?

Yes, based on currently announced plans. Like broadcasters, some cable programmers are planning to offer HDTV programming. In particular, some cable programmers are planning to offer high-definition versions of "premium" content such as movies and live sporting events. If you are considering the purchase of a DTV set, you should know that you initially may not be able to receive premium cable channels in high-definition formats because copyright agreements have not been completed. Specifically, certain kinds of premium content generally will not be available in HDTV formats until methods for preventing the making of unauthorized copies become finalized. Work on these "copy protection" solutions is underway but is not yet complete. You therefore may wish to ask your local cable operator to describe its plans for offering high-definition cable channels. Specific questions you may want to ask include:

- Is your cable system planning to offer any digital cable programming? If so, what programming will be in HDTV formats?

- Will my new DTV set be able to receive HDTV cable programming when it becomes available?

- If not, can my new DTV set be upgraded with new capabilities or special equipment to receive HDTV cable programming when it becomes available?

14. Will I be able to receive satellite TV services, including new HDTV programming, on my DTV set?

Yes, with the proper equipment. Satellite TV providers currently deliver programming service to consumers using separate satellite reception equipment (typically, a set-top satellite receiver and a satellite dish). Some satellite TV providers recently have announced plans to supply high-definition digital programming for DTV sets. New set-top satellite receivers and slightly larger dishes generally will be needed to receive this programming. In addition, some consumer electronics manufacturers have announced plans to provide new digital sets with a built-in satellite TV receiver. If you purchase one of these sets, you may not need a separate set-top device. You may wish to ask your satellite TV provider about its plans for providing DTV programming, including any plans for providing high-definition programming.

15. Where can I get more information about any of these questions and issues?

For more information on these topics, you can e-mail questions to the FCC's Office of Engineering and Technology at oetinfo@fcc.gov. Also, as discussed above, you should contact your local television station, cable provider, satellite service provider, and consumer electronics retailer to learn more about specific issues. The FCC also will provide additional consumer bulletins in the future to keep consumers informed as the DTV transition progresses.

Chronology

We live in a world of electronic marvels today that entertain us with pictures, sound, and music, bring the entire world into our living rooms, allow us to communicate over vast distances even without wires, provide us with information processing and computing capability of unimaginable power, control our machines, and give us a host of other functions that enhance our lives. And for the most part, we take it for granted.

But these things did not happen overnight. They're the culmination of discoveries and inventions from the minds of great scientists and tinkerers over nearly the past two centuries. This chronology, developed by the Consumer Electronics Manufacturers Association, provides an overview of the key developments, discoveries, and inventions during that time, that have brought the world to its current state of technological advancement.

1822: Charles Babbage develops the Difference Engine, the first—albeit enormous—calculator.

1832: Babbage conceives the first computer, the Analytical Engine, a mechanical calculating machine driven by external instructions akin to modern-day software. He never builds it.

1840: Alexander Bain proposes a fax machine that uses synchronized pendulums to scan an image at the transmitting end and send electrical impulses to a matching pendulum at the receiving end to reconstruct the image. The device, however, is never practically developed.

1844: Samuel F.B. Morse sends telegraph message, "What hath God wrought?"

1863: Giovanni Caselli receives U.S. patent for fax machine, based on Bain's ideas, called the "pantelegraph." Service between Paris and Lyons, France, operates between 1865-70, but the Franco-Prussian War ends the experiment.

1872: Thomas Edison patents an electronic typewriter, the prototype for later teletype machines.

1876: Alexander Graham Bell receives U.S. patent for telephone.

1877: Thomas Edison invents the phonograph.

1882: Lars Ericsson invents the telephone handset with microphone and earpiece combined.

1883: Thomas Edison patents the Edison Effect, which later is developed into the vacuum tube by Lee deForest.

1884: Paul Nipkow invents scanning disc for first mechanical television.

1888: Emile Berliner applies for patent on flat phonograph disc.

Thomas Edison and William Dickson make a sound "motion picture" with a phonograph synchronized with a Kinetoscope.

Heinrich Hertz detects and produces radio waves.

Oberlin Smith describes a magnetic sound recording system, a forerunner of the magnetic tape recording system.

1893: Valdemar Poulson invents the wire recorder, in use until the magnetic tape recorder is developed just after World War II.

1895: Guglielmo Marconi sends and receives wireless signals in Italy; Alexander S. Popoff accomplishes same feat in Russia.

1897: German scientist Karl Ferdinand Braun constructs first cathode ray tube scanning device, a technology still in use today in TVs and computer monitors.

1900: Professor Reginald A. Fessenden discovers AM radio.

Constantin Perskyi coins the word "television."

Two German scientists, Johann Elster and Hans Geitel, invent the photoelectric cell.

1901: Marconi receives the first transatlantic wireless telegraph message.

1902: Arthur Korn develops the first practical fax machine using selenium cells to deconstruct pictures into components, then reconstructing them at the receiving end.

George Blickensderfer invents the first practical electric typewriter.

1904: John Ambrose Fleming files patent for the vacuum tube.

1906: On Christmas Eve, Fessenden transmits voice and music via AM radio.

1907: Boris Rosing in Russia and A.A. Campbell-Swinton in England simultaneously develop image reproduction methods using electromagnetic scanning.

1912: Institute of Radio Engineers forms.

Lee deForest invents the audion tube.

1913: Edouard Belin invents the modern fax machine, which uses ordinary phone lines.

1915: First transatlantic radiotelephone call; first transcontinental phone call made.

1918: Edwin Armstrong develops the superheterodyne radio receiver, still the basic technology for all radios.

First binary calculating machine.

1919: Shortwave radio developed.

Radio Corporation of America (RCA) founded.

1920: Commercial radio broadcasting begins (KDKA, Pittsburgh); first radio receivers for sale.

1922: John Carson describes concept of FM radio.

Farmboy Philo T. Farnsworth envisions electronic TV system; the resulting sketch later proves Farnsworth's patent claim.

1923: National Association of Broadcasters (NAB) formed.

First transatlantic radio broadcast.

Complete TV system including kinescope, or picture tube, demonstrated by Dr. Vladimir K. Zworkin; Zworkin applies for patent of iconoscope or TV camera tube.

1924: Radio Manufacturers Association, predecessor of EIA, founded.

Loudspeakers replace earphones.

Western Electric Company patents electrical sound recording.

Zenith Electronics Corporation produces first portable radio.

1925: John Logie Baird produces TV picture of human faces.

Vladimir Zworkin files for color TV patent.

1926: Zenith introduces AC radio receivers designed to plug into electrical outlets.

1927: Philo Farnsworth applies for patent on electronic television.

First coast-to-coast radio hookup.

Bell Telephone Laboratories demonstrates wireless TV between Whippany, New Jersey, and New York.

First picturephone conversation.

Boris Rtcheouloff applies for videotape recording patent.

John Logie Baird creates first videodisc.

1928: Federal government issues first experimental TV station permits; General Electric, NBC, and BBC begin TV broadcasts within the next year.

AT&T sends motion pictures from Chicago to New York, the first successful trial of video delivery through telephone lines.

Kenjiro Takayanagi demonstrates cathode ray system in Japan.

1930: I.G. Farben AG develops magnetic plastic audio recording tape in Germany.

1931: First patent for stereo recording issued.

1932: Blaupunkt G.m.b.H. introduces automobile radios.

Marconi discovers microwaves.

1933: Howard Armstrong patents FM radio.

1934: Federal Communications Commission (FCC) established.

1936: Berlin Olympics televised.

1937: Chester Carlson invents xerographic copy method, but it will be 22 years before the first commercial Xerox Corporation photocopy machine is introduced.

John Atanasoff begins work on first electronic computer, completed 1939.

1938: Konrad Zuse invents first binary computer.

1939: Television introduced at New York World's Fair.

RCA, GE, DuMont, Philco, and two other companies sell first television sets.

First experimental FM stations go on air.

1940: First color TV broadcast.

1941: Commercial FM operation begins.

1942: TV equipment production banned for war.

1945: Unlimited output of radios permitted after the end of World War II.

Arthur C. Clarke proposes geosynchronous communications satellite system; area in space where these satellites orbit 20 years later dubbed the "Clarke Belt" in his honor.

Grace Murray Hopper coins the term "bug" to describe a computer fault.

1946: First mobile telephones.

First electronic computer, ENIAC, demonstrated.

John Mullin demonstrates improved German-made Magnetophon magnetic tape recorders for U.S. engineering convention.

1947: William Shockley, Hohn Bardeen, and Walter Brattain invent the transistor at Bell Telephone Laboratories.

1948: 45 rpm and LP records introduced.

First magnetic tape recorders sold in United States by Ampex Corp.

Gone With the Wind transmitted electronically to the Library of Congress in less than 2½ minutes.

TV set sales increase more than 500 percent compared to 1947 sales.

First cable TV systems in United States.

1949: First three-speed phonographs marketed.

First pagers announced.

1951: Jack Mullin, working for Bing Crosby's Labs, demonstrates magnetic videotape recorder, an altered audiotape recorder.

Transcontinental TV inaugurated in United States.

1952: UHF TV broadcasts authorized.

First transistorized device, a hearing aid, is sold.

1953: NTSC-compatible color TV successfully demonstrated to FCC; FCC authorizes broadcasts to begin January 22, 1954.

1954: Color TV broadcasting begins.

First mass-market transistor "pocket radio" introduced at $49.95.

1955: Narinder Kapany introduces optical fiber.

Bell Labs builds first transistorized computer.

1956: Black-and-white portable TV era begins.

Ampex introduces commercial videotape recorder.

First transatlantic telephone cable operational.

Computer hard drive developed.

1957: Radio-Electronics-Television Manufacturers Association changes name to Electronic Industries Association.

1958: Stereo records and phonographs introduced.

Jack Kilby of Texas Instruments and Robert Noyce of Fairchild Semiconductors separately invent the integrated circuit.

FCC creates Citizens Band (CB) radio service.

Modems introduced.

1959: First Xerox copier introduced.

1960: First telephone answering device available to consumers.

First rectangular-screen TV introduced.

First battery-operated transistorized TV for sale.

Echo, first communications satellite, launches.

1961: Stereo incorporated into FM radio.

1962: Legislation passes requiring all-channel tuning (UHF and VHF) in television receivers.

1963: Philips Electronics NV introduces the compact audio cassette.

Optical videodisc, basis for today's laserdisc and DVD, first demonstrated.

1964: AT&T demonstrates futuristic Picturephone at World's Fair.

American Standard Code for Information Interchange (ASCII) adopted.

First commercial device with integrated circuit, a hearing aid, sold.

Sony engineer Koichi Tsunoda proposes videocassette.

1965: Eight-track audio players marketed.

First commercial telecommunications satellite, Early Bird (Intelsat 1), launched.

Sony Corporation introduces first portable consumer video recorder.

1966: Integrated circuits introduced into consumer products, starting with pocket calculators and electronic watches.

1967: First Consumer Electronics Show (CES) held in New York.

Ray Dolby develops noise reduction system.

1968: Carterphone decision allows consumers to connect equipment to phones, allowing manufacturers to market telephone answering devices.

Computer mouse demonstrated by Douglas Engelbart, commercialized by the Apple Macintosh 16 years later.

1969: Four-channel stereo tapes and players marketed.

Arpanet, forerunner of Internet, becomes operational.

1970: Intel Corporation introduces computer memory chip.

Computer floppy disk developed.

1971: Electronic tuning first seen in U.S. TVs.

Consumer electronic calculator introduced.

First dot matrix printers introduced.

1972: First home video games, designed to play through TV receivers, marketed.

Sony introduces U-Matic VCR, forerunner of Beta.

1973: First Winter Consumer Electronics Show held in Chicago.

Giant-screen projection color TVs marketed.

1974: United Nations establishes international fax standard, Group 1.

1975: First Betamax VCR introduced.

First personal computer, Altair 8800, debuts.

CB radio becomes major consumer product.

Laser printer introduced.

First liquid crystal displays (LCD) sold.

1976: First microprocessor video games marketed.

First VHS VCR introduced.

Inkjet printer developed.

1977: Apple Computer Inc. introduces first personal computer.

Home color TV video cameras sold.

1978: Laserdisc players marketed.

Winter Consumer Electronics Show held in Las Vegas for first time.

1979: Portable language translators introduced.

Personal headset audio introduced.

Institute of High Fidelity merges into EIA.

1980: Closed-captioning decoders sold.

Cellular phone service tested.

World standard for optical digital audio compact disc (CD) established.

First portable VCR-camera combinations (camcorders) demonstrated.

Group 3 international fax standard established.

1981: High-speed fax machines marketed.

IBM PC, using Microsoft Corp.'s Disk Operating System (MS-DOS), introduced.

First portable computer sold.

1982: FCC authorizes AM stereo broadcast.

VHS-C Videotape format introduced (first called UCM, then Mini-VHS).

Dolby Laboratories introduces surround sound for home use.

First compact disc (CD) players for sale in United States.

1983: Beta format VCR-camera combinations introduced.

8mm video format standards set.

Hi-Fi VCRs introduced.

Apple Lisa, first computer using graphic user interface (GUI) and mouse, unveiled.

Cellular service introduced.

1984: First working auto CD system debuts at Summer CES.

FCC authorizes multichannel TV sound broadcast; first stereo TV broadcasts begin.

Sales of stereo color TV receivers and adapters begin.

First color TVs with all-digital signal processing circuitry marketed.

CD-ROM introduced.

Courts break up AT&T.

1985: Color TVs with 35-inch picture tubes marketed.

8mm video home VCR decks and software introduced.

Still-picture magnetic disc video recorders, players, and printers demonstrated.

First portable LCD TV sold.

1986: First consumer video telephone marketed.

First digital audiotape recorders demonstrated.

4mm video format for camcorders demonstrated.

Scrambling of satellite-fed cable TV programming starts; sale of decoders and program subscriptions to home dish owners begins.

Stereo sound in television broadcasting available in all major U.S. population centers.

Radio Data Service (RDS) begins in Europe.

1987: Higher resolution VCRs and camcorders introduced (S-VHS and ED-Beta).

Compact Disc Video (CD-V) introduced.

First Advanced Television (ATV) system demonstrated.

Movie theater experience moved into the living room; Dolby Pro Logic is available at home.

1988: First Improved Definition Television (IDTV) receivers marketed.

High-band 8mm camcorder format debuted.

Recordable CDs demonstrated.

CD-Graphics format introduced.

Still video cameras introduced.

First transatlantic fiber-optic cable laid.

1989: First 16-bit video games introduced.

1990: First digital audiotape (DAT) recorders introduced.

Production of giant-screen (27 inches or larger) color TV picture tubes starts in United States.

Legislation requiring closed-captioning decoders in all larger color TVs manufactured after July 1, 1993, signed into law.

All-digital high-definition television (HDTV) systems proposed; FCC sets field testing schedule.

Color fax machines sold.

1991: Multimedia CD-ROM computer systems introduced.

First TVs with built-in closed-caption display capability introduced in United States.

U.S. testing of HDTV systems begins.

EIA and National Association of Broadcasters (NAB) set AMAX and AMAX Stereo standards for advanced AM radio performance.

1992: Digital cellular phone service introduced.

Digital Compact Cassette (DCC) first marketed.

MiniDisc first marketed.

Radio Broadcast Data Service (RDS) introduced in United States.

CEBus Home Automation standard finalized in United States.

Color picture videophone available for home market with $1,500 price tag.

Mobile Electronics Association merges into EIA.

1993: Grand Alliance for development of HDTV system formed.

16:9 Aspect Ratio (widescreen) television sets marketed in United States.

Next-generation video games announced.

Personal Digital Assistants (PDA) introduced.

FCC adopts signal standard for AM stereo broadcasting.

Video CD adopted as a five-inch optical disc standard.

First TV sets equipped with VCR Plus+ debut in Japan.

First 625-line chip-based 16:9 widescreen format camera introduced.

First wireless headset portable CD player marketed in Japan.

900 MHz cordless phones introduced.

1994: GPS Auto Navigation systems marketed in United States.

Direct Broadcast System (DBS) dishes launch.

Pentium and PowerPC microprocessor chips introduced.

First CES Mexico held in Mexico City.

1995: Competing DVD standards introduced; single DVD standard selected.

First television program (*Computer Chronicles*) delivered via the Internet.

Internet and commercial on-line usage explode; consumer on-line segment jumps 64 percent to almost 15 million users while about 8 percent of U.S. households are hooked into the Internet.

First television station KOLD 13 uses a networked digital video server in its daily on-air operations.

Sony announces first digital camcorders to be sold worldwide.

EIA's Consumer Electronics Group (CEG) becomes CEMA, Consumer Electronics Manufacturers Association.

Interactive cable modem trials with consumers started.

Quad-speed CD-ROM drives become common feature of multimedia PCs.

Flash memory technology standard introduced. FCC approves 15 additional channels for cordless phones; 25-channel cordless phone models introduced.

RDS pagers introduced.

Safety Warning System (SWS) technology standard finalized.

Flat-screen plasma display TVs introduced.

First GPS-based vehicle recovery systems marketed.

First voice-recognition car security systems marketed.

Multifunction home office products become popular.

Digital Satellite System (DSS) installations reach 1 million mark.

1996: First TV sets equipped with VCR Plus+ introduced in United States.

HDTV is broadcast and received live at commercial station WHD-TV in Washington, D.C.; FCC accepts Grand Alliance standard.

Set-top boxes plug into TV and allow viewers to surf the Internet's World Wide Web via remote control.

Congress passes Telecommunications Act of 1996, containing a directive to television manufacturers to begin installing "V-Chips" in TVs that would allow selected programming to be blocked.

Zenith introduces the U.S. market's first HDTV-compatible front-projection TV.

First DVD players sold in Japan.

Agreement between broadcasters, TV manufacturers, and PC makers sets interindustry standard for HDTV. HDTV set to hit shelves in 1998.

1997: Smart phones allow access to the World Wide Web and e-mail via a phone with a screen and keyboard.

Debut of new Intel Pentium processors with MMX technology said to boost PC's multimedia performance.

56 kbps modems hit the market.

DVD players debut in United States.

First two-way pagers.

INDUSTRY ORIGINS AND HISTORY

Even with a chronology such as the foregoing, it's sometimes difficult to comprehend the degree to which consumer electronics has changed in the last several decades, and just what magnitude of change it has made in our lives. For example, here's an interesting factoid: a 21-year-old today was just about four years old when compact disks were introduced. So, if it seems that many of today's younger people are unfamiliar with vinyl long-playing records, it should not be surprising. And rental of movie videotapes became popular at about the same time, so many people of age 20 or less do not remember a time when Hollywood movies couldn't be viewed in the comfort of their own homes.

The following synopsis of the origins and history, developed by the Consumer Electronics Manufacturers Association, helps to put the current state of development of consumer electronics into perspective.

From a technological point of view, the recent nostalgia for the 1970s in film, music, and fashion is kind of curious. If sent back in time, today's primary filmgoers, music buyers, and fashion pacesetters wouldn't recognize the world into which they were born.

Most, if not all, of the leisure conveniences that we blithely take for granted had yet to be invented or developed when today's 20-somethings were born. No one had personal stereos to clip to belts or toss into backpacks. If they wanted to watch a film, they actually had to go to a theater; there were no VCRs. To listen to the latest hits, they'd have to buy scratchy and fragile vinyl albums or bulky 8-track tapes; there were no CDs. School reports had to be written out in longhand or typed on a typewriter, and all research had to be done in a library; the personal computer was a decade away and the Internet was still a Department of Defense experiment. "Big-screen TV" was 25 inches. Cable was new and reached only a small percentage of American homes. Video games consisted of two "paddles" that were actually just lines on opposite sides of a TV screen that could be moved only up or down to deflect a square "ball." If you wanted to make a phone call away from home, you had to use a pay phone; only spies and chauffeured executives had car "radio" phones. If you wanted to capture images of important events in your life, you used film, either for static stills or grainy Super 8 movies. If you had a job, you went to an office where all the massive and expensive office equipment was located; SoHo was still just a section of New York City and London.

There is no doubt that the last quarter century has been the greatest period of scientific and technological invention and development in world history. The foundations for the technological wonder world of the next millennium often envisioned by futurists, philosophers, writers, and filmmakers in decades and centuries past have been laid in the last 25 years.

While we may not have regular passenger service to the moon as Arthur C. Clarke predicted in *2001: A Space Odyssey*, his 1945 vision of a string of satellites in geosynchronous orbit delivering global communications was so dead-on that the orbital satellite parking lot 22,000 miles above the equator was named the Clarke Belt in his honor.

The apex of this era of invention and development may have taken place on Friday, February 27, 1998, when WFAA-TV in Dallas, Texas, began the first commercial HDTV broadcasts. Never mind that there were no digital television sets yet in consumers' homes to receive the signal, and never mind that the transmission temporarily shut down wireless heart monitors at a nearby hospital. Future historians may point to WFAA's HDTV telecast as both the technological curtain call of the 20th century and the inaugural act of the 21st century's world of tomorrow.

What is more amazing is how nonchalantly we have accepted and absorbed this technological new world order into our daily lives. While sociologists and theologians debate how—or if—these new technologies have changed basic human nature, there is no doubt that they have changed human behavior. Products that have debuted in the last 25 years—the compact disc, the VCR, the camcorder, CB radio, the personal computer, video games, cordless and cell phones, pagers, satellite TV, and home theater—have affected the lives of every American. By the end of the century, we may be able to add DVD and HDTV to that list.

The Foundations

If America of the 1970s suddenly seems primitive, consider America of the 1870s. When America celebrated its centennial, electricity was either a phenomenon encountered in the winter because of charged clothing, or the product of a thunderstorm. A "consumer electronics product" was no more than a fantasy.

Some say that the modern consumer electronics industry was born with the commercialization of radio in the early 1920s. But it was during the 50 years straddling the 19th and 20th centuries that almost every major consumer technology we enjoy today originally was developed, usually by men working on their own, following a vision only they believed.

The first of these men was Thomas Edison. The year after America's centennial, Edison became the most famous and admired man in America. In December 1877, he sketched a diagram of a strange device and asked one of his technicians in his Menlo Park, New Jersey, lab to build it. When the technician brought back the machine, he asked his boss what it was. Edison adjusted the great horn that had a needle attached to a wax cylinder covered with tin foil. Turning a crank on the side, he shouted into the horn: "Mary had a little lamb, its fleece was white as snow, and everywhere that Mary went, that lamb was sure to go."

He then readjusted the horn and its needle, which had just scratched a path on the tin foil, then turned the crank again. To the amazement of the assembled technicians, out of the horn came Edison's words precisely as he had spoken them. Thus was born the phonograph.

The phonograph, however, initially languished as a consumer product. It wasn't until almost 20 years later that the phonograph made it into consumers' homes, and it wasn't Edison who put it there. A German-born telephone expert named Emile Berliner flattened out Edison's wax cylinder and created a disc, then figured out a way to motorize the whole operation. Gramophones—the first record players—were sold starting in 1896 by the newly formed Victor Talking Machine Company. In 1907, the Gramophone as furniture—the Victrola—was introduced, and great music was now available to the masses.

But the phonograph was not the most important of Edison's discoveries in modern consumer electronics. His most important contributions are the central power station and something called "The Edison Effect." The first thing you do when you buy a stereo, TV, or any electronic gizmo is plug it into a wall socket. This simple act is possible because Edison spent almost his entire fortune and staked his substantial reputation to build the Pearl Street power station in lower Manhattan in 1882. When Edison gave the order to flip the switch on Monday afternoon, September 4, 1882, stores all along Fulton and Nassau streets, the editorial offices of *The New York Times*, and the brokerage house of Drexel Morgan were lit up with his incandescent light bulbs that his power station fueled. The world had changed forever.

But Edison was producing direct current, dangerous and difficult to transmit efficiently over long distances because it flowed only one way. After a bitter fight between Edison and Westinghouse, the system now called alternating current (AC), which allowed electricity to flow bi-directionally, reigned supreme. AC was invented by a nearly penniless Croatian immigrant named Nikola Tesla, recently arrived in America after working in the Paris offices of the Continental Edison Co. (ConEd). This form of electrical current remains the standard for transmitting electricity from power plants to homes and businesses.

The Edison Effect was an accidental by-product of Edison's invention of the light bulb in 1879. In 1883, while tinkering with ways to make his bulb last longer, he discovered that the flow of electrons inside a vacuum could be controlled by electronic and magnetic fields. Since Edison was essentially a really imaginative mechanic, and not a university-educated scientist, he didn't quite envision the practical application of the discovery. He simply patented it and forgot about it. In fact, what Edison had discovered was a precursor to the vacuum tube, the basis of every piece of electronics invented during the next 65 years, and still the basis of the cathode ray tube used in TVs and computer monitors.

The reason Edison's electron tube was useless was that no one knew radio waves existed. In 1872, Scottish physicist John Clerk Maxwell theorized that electromagnetic waves existed, but it wasn't until 1888 that German physicist Heinrich Hertz first detected and produced electromagnetic waves.

Guglielmo Marconi also realized that electromagnetic waves could transmit signals the way a wire carried electricity. In 1896, Marconi went to England and set up the Wireless Telegraph and Signal Company. In St. John's, Newfoundland, Canada, on Dec. 12, 1901, he received the Morse letter "S" transmitted from Poldhu, Cornwall, England, the first wireless signal to cross the Atlantic Ocean.

Radio

Transmitting dots and dashes was, however, quite a long way from transmitting voice. Twenty years after Edison discovered his effect, English scientist John Ambrose Fleming, working for Marconi, put the effect to practical use, inventing the first true diode vacuum tube used in Marconi's first trans-Atlantic transmission. In 1912, American inventor Lee deForest took the concept a step further and created an amplifying vacuum tube he called the audion tube, the essential component in what would become known as radio.

While the development of transmitting technology progressed, Canadian engineer and inventor Dr. Reginald Fessenden was figuring out ways to transmit something more useful than telegraph code. In December 1900, he succeeded in transmitting his voice a mile; by 1904 he had discovered amplitude modulation—AM. In 1906, Fessenden convinced several ships off the coast of Massachusetts to install AM receivers. On Christmas Eve, Fessenden played the violin, read from the Bible, and played Gramophone records to become the first "DJ." His transmissions, received by operators as far away as Virginia, became the first true radio broadcast.

Radio soon became a hobby and obsession. Wireless operators became folk heroes. They included a 21-year-old Marconi telegraph operator named David Sarnoff, who became famous as the man who first broadcast news of the Titanic disaster in April 1912. A year later, 24-year-old Edwin Howard Armstrong patented an improved receiver he had tinkered with while attending Columbia University. In 1918, he invented the superheterodyne radio receiver, the principal of which still is used in every radio device made today.

It was around this time that the word "radio" came into wide usage, derived from its use of radiated signals. American boys everywhere started buying radio kits in much the same way they later would build plastic model airplanes.

Almost anyone could set up a transmitter and send signals, and many did, creating a radio Tower of Babel. In 1919, Assistant Secretary of the Navy Franklin Roosevelt advised that, in the interests of national security, the radio business ought to get organized and standardized. On Oct. 17, 1919, most radio patents voluntarily were transferred to a new company owned by several of the major radio players—the Radio Corporation of America, or RCA. In charge of the commercial side of the new entity was the hero of the Titanic reports, David Sarnoff. Sarnoff was not a scientist but a visionary, and would remain the best known and most influential figure in the world of consumer electronics and communications for nearly 50 years.

In 1920, a Pittsburgh radio amateur and Westinghouse employee named Frank Conrad started broadcasting music. His ad hoc evening programs were so popular that his employers took interest, set up a studio for him in the building and dubbed the operation KDKA, the first commercial radio station. The first commercial radio broadcast was the presidential election returns on Nov. 2, 1920, heard by a few thousand enthusiasts in the Pittsburgh area.

The resulting publicity sparked a rush to start radio stations; by 1923 there were more than 500. One hundred thousand radios were sold in 1922, at an average cost of $50 each. By 1924, the annual factory dollar volume of radios had multiplied tenfold, to $50 million.

Radio became as ubiquitous in American homes as bathtubs. In fact, the consumer electronics business was known as the radio business for decades afterward. Not only were there radio models that required headphones, but there were fancy furniture models, car radios (1932), primitive "boom boxes," and combination radio-phonographs. The golden age of radio spanned a 21-year period, from 1920 until America's entry into World War II. During those years, 100 million radio receivers of increasing sophistication, quality, and size were sold.

Up to this point, all radio broadcasts were on the AM band. In 1922, American mathematician John Carson theorized a higher quality frequency modulated, or FM, scheme. In 1933, Armstrong patented the static-free radio technology, but because FM required an entirely new infrastructure and equipment for both broadcast and receiving, few radio companies were interested. After a brief fling at RCA in the mid-1930s, Armstrong was forced to create FM himself, founding the first FM radio station at his alma mater, Columbia University, in 1941. Armstrong left RCA because Sarnoff had focused his attention on television.

Television

It is generally believed that television is a technology of the 1950s, springing fully formed like Athena from the head of Zeus in the gray-flannel period of prosperity following World War II. But while television exploded in those years, it was actually an old idea—or rather, a dream—that survived a long and relatively violent birth. Once radio became a familiar technology, it was a natural step to imagine sending pictures as well as sound through the air. In 1884, 24-year-old German engineer Paul Nipkow started playing with a spinning wheel with perforated holes that a thin beam of light passed through to scan an image that could be transmitted. The Nipkow Wheel was the first example of mechanical television technology, and for 40 years it was the only television technology.

In 1922, however, a 15-year-old Idaho farm boy fascinated by radio technology, Philo T. Farnsworth, made a sketch of a system in which a "gun" shot electrons onto the inside of a cathode ray tube. That sketch would later prove precedence for Farnsworth's television patent claims.

Advancements in television seemed to come in bunches. Farnsworth was not alone in thinking up a television scheme. In 1925, John Logie Baird, a Scottish engineer, produced the image of a recognizable human face and actually got his Nipkow-based system temporarily accepted by the BBC after a series of successful public demonstrations. And Russian émigré and Westinghouse employee Dr. Vladimir Zworkin filed for a patent on an electronic color TV system almost identical to Farnsworth's. By 1928, several companies, including NBC and the BBC, started broadcasting television pictures, each picking its own favorite technology.

It was Farnsworth's and Zworkin's almost identical systems, however, that seemed the most intriguing. By the late 1920s, Zworkin was developing his system for Westinghouse, and then RCA, and Farnsworth was working in San Francisco on behalf of private backers, each racing to perfect the necessary technologies. A long legal and business battle followed to see who owned the patents—a battle Farnsworth ultimately won. But it is Zworkin who generally is cited as the father of television. Even though the BBC began the first regular television broadcasting in 1936, and Farnsworth actually had a working TV studio in a Philadelphia suburb that same year, it was Sarnoff who initiated the age of television. At the New York World's Fair in April 1939, NBC began the first regular broadcasts in the United States, starting with the address by President Franklin Roosevelt officially opening the exhibition.

A year later, Peter Goldmark made the first color TV broadcast for CBS. In 1945, the FCC designated 13 channels (later 12) for broadcasting VHF signals, and channels 14 to 83 for UHF signals, although the first UHF stations didn't start broadcasting until 1952. The standard for color TV was set in 1953, and Zenith Electronics Corp. conceived an era of couch potatoes when it perfected the first practical remote control in 1956. The first all-channel—VHF and UHF—sets were mandated in 1964.

World War II interrupted further commercialization of broadcasting of any kind, as all civilian production of television and radio sets was suspended during the war. But postwar prosperity would see the real growth in not only these industries, but the entire consumer electronics industry.

Color TV

The war might have ended in 1945, but a nasty battle raged for a decade afterward over color television. But first, TV had to fight off radio.

By 1947, only 250,000 TV sets were in use and in 1949, radio boasted 81 percent of all broadcast audiences. By the end of that year, however, TV claimed 41 percent. By the end of the following year, the average daily radio listening session was down from nearly four hours to 24 minutes. By the end of 1951, there were nearly 8 million TV sets in use. It was a short fight.

Color had been the Holy Grail of television almost from its infancy. As early as 1929, Bell Labs had demonstrated color TV transmissions. The original intention was to create a color system compatible with existing black-and-white sets, but this soon proved technologically impossible. All through the war, CBS and NBC were experimenting with competing color systems.

CBS used a mechanical system based on the Nipkow wheel and developed by Peter Goldmark. The so-called "compatible" RCA system was an electronic system based on Zworkin's and Farnsworth's work.

At first, the FCC couldn't make up its mind, but in 1950 it approved CBS's system. The National Television Standards Committee (NTSC), formed in 1950 by dozens of companies with stakes in the battle, investigated, formed subcommittees and held hearings. In mid-1953, the NTSC recommended to the FCC that it reverse its earlier decision. In December 1953, the FCC did, approving the all-electronic NTSC system that is still the standard today.

Sales of color TVs, however, were slow. *Time* proclaimed that color TV was "the most resounding industrial flop of 1956," but in the fall of 1960, Sarnoff gave people a reason to buy them. His NBC began airing *Walt Disney's Wonderful World of Color* and color TVs began flying out of the stores.

The Age of Miniaturization

The war prompted research and development of advanced technologies on an unprecedented scale. Atomic bomb research in New Mexico spurred the development of the computer. Other technologies were refined and perfected far faster than would have been possible during peacetime. These technologies, converted into commercial products after the war, included radar, lasers, plastics, microwaves (discovered by Marconi in 1932), and satellite communications.

The most important of these developments was the transistor. In 1939, William Shockley, a researcher at Bell Labs, wrote in his notebook that he thought it was possible to replace the clunky vacuum tube with semiconductors. By December 1947, Shockley, John Bardeen, and Walter Brattain had perfected the transistor, inaugurating the age of miniaturization.

AT&T licensed the technology cheaply to other manufacturers. In 1952, AT&T waived all patent rights for the transistor's use in hearing aids, the first device to use transistors. The hearing aid was followed two years later by the first transistor radio, the Regency TR-1, which retailed for $49.95. The Regency was not the first "pocket" radio, but because of its transistors, it lasted more than 30 hours on a single battery, as opposed to the mere three to five hours of older tube models.

Three years later, a small Japanese company rising from the ashes of the war, Tokyo Telecommunications Engineering, introduced a smaller model, the TR-55, under its new brand name - Sony Corp. Sony's success spurred other Japanese companies and, by the end of 1957, three dozen Japanese companies had exported more than 100,000 transistor radios to the United States. In 1959, the United States imported more than 6 million, half of Japan's total output of transistor radios, which accounted for half of Japan's source of U.S. dollars and the country's fourth largest export. What drove these sales was not necessarily the technology, but rock 'n' roll. The primary buyers of transistor radios were teenagers.

Things got smaller in 1958. Jack Kilby, working for Texas Instruments, and Robert Noyce of Fairchild Semiconductors separately invented the integrated circuit, the basis for all computerized devices.

Hi-Fi

To hear music before Edison, you had to go to a concert. The Gramophone brought music home, as did the radio, but the sound quality was nothing like a concert experience. After the war, quality music reproduction became the province of the home hobbyist, thanks to the development of high-fidelity stereophonic recording and playback, otherwise known as hi-fi. Hi-fi was actually the convergence of several technologies. The first was the concept of stereo—sound from multiple speakers that simulated a live performance. The first patents for stereophonic recording and reproduction were issued in 1931, but technology at the time—both on the recording side and the receiving and playback side—was ill-equipped to handle the actual stereo signals.

The transistor was the first breakthrough, creating more efficient signal amplification. Next was the refinement of the physical playback media. In 1930, the German industrial giant I.G. Farben AG developed the first magnetic tape recorder, and by 1935 Adolf Hitler was using magnetic tape to record and broadcast speeches.

By the end of the war, rumors were pervasive in the Army Signal Corps that the Germans had developed high-fidelity tape recorders. Corpsmen scoured the French and German countryside, procuring the remnants of the German recording and radio industry. Major John T. Mullin shipped one such remnant, a tape recorder called a Magnetophon, and several reels of tape, back to his San Francisco home and started to tinker.

In May 1946, Mullin presented his rebuilt tape recorder to a radio industry convention. Bing Crosby's engineers quickly hired Mullin and his machine to tape the singer's radio show. Two years later, a tiny company called Ampex Corp., with Mullin's help, introduced the first commercial tape recorder in the free world, making high-fidelity stereo recording possible.

That same year, Hungarian refugee physicist Peter Goldmark, the same scientist who developed CBS's color TV system, developed the 33-rpm long-playing vinyl record, a disc with much longer playing time and much higher fidelity than the 78-rpm wax discs that were then the norm. The following year, RCA introduced the 45-rpm disc, the so-called seven-inch "single" that would become the primary method for disseminating rock 'n' roll music.

Most early hi-fi recordings were limited to reel-to-reel tape. High-fidelity stereo finally came to vinyl records in 1958, but even then not all recordings were in stereo. It wasn't until 1963 that Philips Electronics NV developed the audiocassette, which made listening to tape much easier, thanks in part to a new noise reduction technology developed by Ray Dolby in 1967 that removed background tape hiss. In 1961, Zenith brought hi-fi stereo to FM.

Hackers, nerds, and other assorted engineers began to cobble together amplifiers and stereo speakers to play back these new media at home for thousands of hi-fi enthusiasts. In the late 1940s and throughout the 1950s, small companies founded by such entrepreneurs as Avery Fisher, Jim Lansing, Sidney Harman, Hermon Scott, Henry Kloss, Amar Bose, and Saul Marantz sprang up to manufacture hi-fi equipment. Meanwhile, several retailers and mail-order houses, primarily in New York and Chicago, sold the new gear.

The Video Age

Social issues and the moon race took precedence over consumer technology development during the turbulent 1960s. But as the last strains of Watergate and the first notes of disco were heard, corporate engineers and physicists in Japan, and math nerds in the San Francisco Bay area, were marching to much different drummers. The effects on daily life of the two technologies these disparate group of geniuses were conspiring to create—home video and the personal computer—would be as profound as the telephone and television had been in their own time.

While the very public battle over color TV was being waged, a smaller skirmish was being fought in three labs—RCA; Bing Crosby's Labs, headed by audiotape pioneer John Mullin; and Ampex—to develop videotape recording. In April 1956, a six-man development team at Ampex headed by Charles Ginsburg and including a young college student named Ray Dolby, shocked the National Association of Broadcasters convention in Chicago with the first videotape recording.

During the next 20 years, several companies attempted to bring some kind of video recording system into homes, but consumers were resistant to reel-to-reel and film-based solutions.

The idea of a videocassette first was proposed by Koichi Tsunoda, a Sony engineer, in 1964, after seeing Philips' audiocassette. But the question was whether consumers were interested in recording TV programs. There were also myriad legal issues to resolve—was it legal to record copyrighted programs off the air? In spite of these questions, several companies raced to bring some sort of home video recording system to market. It was clear by the mid-1970s that videocassettes were the way to bring video recording to the home. But there were two companies perfecting incompatible formats: Sony, with its Beta system, and JVC, with its Video Home System, otherwise known as VHS, developed by chief engineers Yuma Shiraishi and Shizuo Takano under the tutelage of cathode ray tube pioneer Dr. Kenjiro Takayanagi. Konosuke Matsushita, founder of Matsushita, JVC's parent company, tried to mediate and get the two companies to agree on a single format, but failed.

So, in February 1975, Sony introduced the first Betamax in the United States, telling consumers, "Now you don't have to miss *Kojak* because you're watching *Columbo* (or vice versa)." And two movie studios, Disney and Universal, sued, claiming that home recording constituted copyright infringement.

Two years after the Betamax was introduced, RCA, in October 1977, unveiled its Matsushita-made VHS SelectaVision VCR. RCA marketing executive Jack Sauter made sure that each machine was packed with prerecorded tapes, including one of Muhammad Ali's greatest fights, and backed the introduction with a $4 million advertising campaign. In six months, the VCR format war was practically over. The VHS machines, which could record for four hours—enough for an entire football game—caught and passed Beta as the video-recording format of choice. By the summer of 1979, VHS was outselling Beta by a margin of two to one.

The inventors of the VCR believed that time shifting would drive VCR sales. But the law of unintended consequence imposed itself. No one envisioned the flashing of "12:00" on the all-important VCR clock, nor the demand for renting and watching movies at home.

As the VCR became more important to the consuming public, both the legislative and judicial branches, as well as the Hollywood establishment that fought it, bowed to its inevitable benefits. The Home Recording Rights Coalition ensured the legality and acceptance of the new technology. And Hollywood studios established home video divisions to reap the profits from a technology it once considered a threat.

During the next five years, advancements such as front loading, remote controls, and hi-fi stereo were added. In the early 1980s came the camcorder, which put a TV studio into the hands of the average consumer and brought cute babies and accident-prone adults to a top 20 TV show.

Born in the '70s

While the VCR was staking out a substantial claim for itself on the home entertainment landscape, three other technologies also were establishing themselves: Citizens band radio, the video game, and the personal headphone stereo.

Citizens band, or CB as it quickly became known, was only about a year old when C.W. McCall hit the charts in 1976 with his song *Convoy*, which celebrated this versatile two-way communications device. CB, however, was simply an old product in new clothing. It initially was introduced in 1958 as a two-way communications service available to everyone, but it took the gasoline shortage and truckers' strike of 1974 to bring CB to the public's attention. It became the biggest electronics fad of the decade in short order.

Many Americans still communicate via CB in their homes, cars, and boats. In 1976, sales hovered around 11 million units, but after the FCC expanded the band to 40 channels and banned the sale of 23-channel units after December 21, 1977, sales declined.

Today, CB is still popular, especially in the form of hand-held, battery-operated units for use in emergency situations or during sports outings. Person-to-person communication has been enhanced with the 1996 allocation of Family Radio Service frequencies, 14 channels and 38 subchannels within the 460 MHz band that allow walkie-talkie-style communication for users within a mile of each other.

Atari Inc. and Nolan Bushnell became household names in 1972 with the introduction of "Pong," the first video game system and game. The video game gave viewers a chance to interact with their televisions. The video game industry quickly became the darling of the adolescent set, offering fast game play without having to go to an arcade. These game systems quickly evolved from the simple "Pong" into complex computers, more powerful than the first home computers of just 15 years ago, and a billion-dollar business.

The audiocassette always had seemed a poor cousin to the vinyl LP—until the cassette player left the house. First cassette decks made it into cars and, eventually, pockets with the introduction of the Sony Walkman in 1979, completely changing the way America commuted. Audiocassettes quickly overtook vinyl as the leading prerecorded music format. By 1983, the audiocassette was the leading format for prerecorded music, according to the Recording Industry Association of America (RIAA), and would stay that way for nearly a decade. But soon, even the audiocassette would be overtaken by a new technology.

Personal computing computers had been in development since the 19th century, when Charles Babbage was trying to perfect his impossibly huge, all-mechanical "difference engine." World War II accelerated the development of an electronic version when fast-computing machines were needed to help solve the complex equations created by the development of the atomic and hydrogen bombs. England's top-secret Colossus, designed by a team headed by Alan Turing and used to decipher German codes, used 1,500 vacuum tubes and is considered the first all-electronic calculating machine. In 1944, Hungarian-born Princeton mathematician John von Neumann conceived stored programming, otherwise known as software. Throughout the 1950s, large computers became omnipresent throughout the military and business worlds, but no one figured anyone would want one at home.

In 1971, Texas Instruments had unveiled the first consumer pocket calculators. By then, a small company called Intel Corp. had invented the first microprocessor memory chip and, by 1973, had progressed to the 8080 chip. Featured on the cover of the January 1975 issue of *Popular Electronics* was the first personal computer, the Altair 8800. Using the Intel 8080 chip, the Altair was invented by an ex-Air Force officer from Georgia, Ed Roberts, and manufactured by his Albuquerque, New Mexico, company, MITS. Nerds could buy these small personal computers in kit form, but they had limited practical applications. They had to be programmed by toggle switches and had no monitors—just a lot of flashing lights. A few years later, Paul Allen and Bill Gates set up shop in Albuquerque to write programs for this new machine.

It soon seemed that every nerd with a garage was building some sort of a personal computer. In 1977, two members of the Stanford Homebrew Computer Club, Steven Wozniak and Steven Jobs, formed a company called Apple Computer Inc., and introduced the first fully assembled personal computer, the Apple II. Within a year they were unable to keep up with the demand.

By 1981, computing giant IBM Corp. saw that there was indeed a market for personal computers and introduced the competing IBM PC. It used an operating system called MS-DOS, provided by Gates and Allen, who had founded a company called Microsoft Corp.

Computers, however, required users to memorize a series of complex commands to operate their machines. In 1984, Apple unveiled the Lisa, which used a graphical user interface (GUI), developed originally at Xerox Corp.'s Palo Alto Research Center (PARC). Instead of a keyboard and commands, a device called a mouse moved the cursor around the screen, now filled with icons and pull-down menus that represented functions and commands. In 1984, this point-and-click technology came to the masses via the Apple Macintosh. A year later, Microsoft brought the GUI to IBM-compatible machines with the introduction of the Windows operating system.

The use of the personal computer and digital networks allowed people to "telecommute"— work at home as efficiently as they worked in the office. A new type of small office/home office sprang into being, with tens of millions of workers abandoning their commute to work from home. Manufacturers of office equipment began designing phone systems, fax machines, and photocopiers specifically for these smaller operations.

Personal Communications

The telephone also was feeling the effects of the digital age. AT&T introduced TouchTone service, which enabled calls to be switched digitally. And when the courts split up Ma Bell in 1984, you could buy your own phone instead of renting it from the phone company, making Alexander Graham Bell's invention a consumer electronics product more than 100 years after it was invented.

In 1997, the first "smart," or screen, phones became available. Similar in concept to the ubiquitous French Minitel phones, screen phones included memory for storage of hundreds of speed dial numbers, as well as access to online information services and access to the Internet without a PC.

Americans first were freed from their phones in the 1970s with the introduction of the first cordless models. But in the mid-1990s, an industrial technology called Spread Spectrum was adapted for consumer use. Spread Spectrum, combined with the opening of frequencies in the 900 MHz range, increased the range of cordless phones tenfold.

Mobile phone service, using radio frequencies, began in 1946. In 1983, the first regular U.S. cellular phone system using microwave transmissions went into operation after two years of testing in Chicago. The first "portable" phones, however, were the size and weight of small bricks and had to be carried in briefcases. One million customers were predicted to use them by 2000. At the end of 1997, nearly 40 percent of all U.S. households had one, with service available in almost every desolate nook and cranny of the country.

In 1992, digital cellular systems that tripled capacity and vastly improved sound quality were introduced. By 1996, cellular phones became pocket-sized and every major carrier was offering digital PCS service, including a raft of messaging and information services and features.

But phones were not the only way people communicated. In 1921, the Detroit police department started alerting its officers simply by transmitting a signal that made a device "beep." Although the government and military started using "beepers" during the 1930s, the first consumer paging systems didn't appear until the mid-1970s.

By the end of 1998, there will be 50 million paging subscribers, including more than just doctors, traveling salespeople, maintenance workers, and emergency personnel, who were the original customers for pagers. Many busy parents carry them to keep in touch with their children, for instance.

In the mid-1990s, text messages dictated to operators could be received and in 1997, paging went two-way. Pagers also got smaller; two companies, Seiko Communications and MTX, a partnership between Motorola and Timex, introduced watches that incorporated pagers.

Home Theater

By 1981, the LP was more than 30 years old, the phonograph more than 100 years old. Sony, Philips, and PolyGram had collaborated on a new format that would offer unparalleled sound reproduction—the compact disc. Instead of mechanical analog recording, the new discs were digital; the music was encoded in binary code onto a five-inch disc that was almost impervious to harm.

Unlike fragile vinyl records, the CD would not deteriorate with continued play, was much more difficult to damage through everyday handling, and held twice as much music, all without having to flip it over. The CD was an immediate sensation when it was introduced to the public in 1982. In 1988, it surpassed vinyl as the home playback medium of choice, then caught and passed the prerecorded cassette in 1996.

At the same time the CD was being developed, so was its sister product, the laserdisc. The original development work was done in the early 1960s, but it took 20 years for laserdisc titles and players to make it to stores.

Once renting movies on videotape became a common activity in the early 1980s, consumers started hungering for ways to replicate the movie experience in the home. To answer the demand, TV manufacturers began to produce TVs with larger screens. This led to the perfection and increased sales of front- and rear-projection systems that projected pictures as large as 10 feet high. In the last five years alone, sales of projection TVs have more than doubled.

Bigger sound also was needed. In July 1983, JVC introduced VHS-HiFi, bringing the stereo sound from the theater to the home. But stereo wasn't adequate to transfer the increasingly loud and immersive movie theater experience. In 1982, a four-speaker scheme called surround sound, consisting of two speakers in front and two behind the movie watcher, was developed by Dolby Labs. Dolby surround sound was expanded to five speakers with Dolby Pro Logic in 1987, then to Dolby Digital in 1995.

The digital video revolution also has reached skyward. In early 1994, digital satellite broadcasting began with satellite dishes measuring just 18 inches in diameter—about the size of a large pizza. Within a year, thanks to the demonstrable benefits of its clear digital pictures, digital sound, and a wide variety of programming, the new satellite systems had sold more units in a shorter period of time than any other new technology.

Hollywood finally realized that there was a great deal of money to be made in home video. Remembering the debacle of the VCR wars, several studios worked with a group of hardware manufacturers on the ultimate home video format. In the spring of 1997, Digital Versatile (a.k.a. Video) Disc (DVD) players and software hit the market. In its first year of introduction, DVD already has passed beyond the laserdisc hobbyist into the mainstream.

The lessons of that DVD alliance provided the blueprint for the next great television advance.

The Grand Alliance

No matter how large a TV picture got, or how digitally perfect the delivery method, the NTSC interlaced standard could never look as good as film. What was needed was a higher definition picture that rivaled celluloid.

The first step was taken in 1987 when NHK, Japan's national television network, presented a highly publicized demonstration of its MUSE, an analog high definition television system, in Washington, D.C. In response, the FCC created an Advisory Committee on Advanced Television Service (ACATS) to spur development of an American HDTV standard. The committee declared an open competition among American companies to create a competing —and superior—American analog HDTV system.

A subsidiary of cable box and satellite converter maker General Instrument Corp., VideoCipher, and its wunderkind engineer Woo Paik, did ACATS one better by coming up with a digital HDTV proposal in 1991.

It soon became clear that one company would be unable to tackle the massive digital HDTV problem. Partnerships between potential competitors were formed. While skirmishes were avoided, ACATS soon could smell major war brewing between the new allied groups.

In May 1993, ACATS Chair Wiley forged all the major HDTV players—General Instrument, Lucent Technologies Inc., the Massachusetts Institute of Technology, Philips North America, Sarnoff Corp., Thomson Consumer Electronics, and Zenith—into a Grand Alliance. The idea: take the best ideas from the best minds from the best companies to create the best HDTV solution.

During the next two years, the technology developed by the Grand Alliance was integrated and tested extensively in independent laboratory and field tests. ACATS unanimously recommended adoption of the proposed Grand Alliance standard in November 1995, followed by the FCC's blessings on Christmas Eve 1996.

The Future . . .

. . . could be seen in full digital wonder at the January 1998 International Consumer Electronics Show. Nearly every major television manufacturer had demonstrations of HDTV, in all varieties and sizes, including the kind of wall TV Ray Bradbury dreamed up in *Fahrenheit 451*. The first HDTV sets were due to be available to consumers in the fall of 1998.

Ironically, the two very different technologies—video and personal computing—that initiated this last quarter-century of invention and development have been courting each other. Thanks to the Grand Alliance's flexible HDTV solution, the marriage between the TV and the personal computer could be consummated in time to usher in the new millennium. If this "convergence" is done right, the couch potato soon may become a jumping bean.

The new millennium also will accompany a new era in personal communications. In the next year, a handful of companies will launch satellites to bring wireless satellite phones to the masses.

But as we move toward a new century, we have become a bit technologically jaded. It seems each day brings a new technological wonder. But many question what utopia—or dystopia— the new technology will lead to: the optimistic future of *Star Trek* and H.G. Wells, or the cynical worlds envisioned by George Orwell and Aldous Huxley.

Fortunately, the efforts of the innovators of the last century—and especially the last 25 years—have bequeathed us a well-lit path to the undiscovered country ahead.

THE HISTORY AND TECHNOLOGY OF TELEVISION

The television sets of today, with their sharp pictures, bright color, 50-plus-inch diagonal screens, and high fidelity surround sound are a far cry from the sets that manufacturers produced during the infancy of television. Those sets had screens that were only a few inches across, monochrome-only picture, and sound systems capable of only a few kiloHertz. For someone who is trying to comprehend HDTV, it's instructive to go back and consider how far we've come. The following examination of the history and technology of television, developed by the Consumer Electronics Manufacturers Association, helps put the long history of the development of television into perspective.

Today's home theaters are based on technologies that were introduced six decades ago. Tiny black-and-white picture tubes have grown to giant, full-color projection systems. Audio has advanced from one tiny speaker to booming multichannel surround sound. And other components have brought movies out of the theaters into millions of homes.

The dream of extending human vision went through two phases: mechanical and electronic scanning. In 1883, Paul Nipkow, a German scientist, invented a mechanical scanning device that could break down an image into a sequence of tiny pictorial elements. His scanning device was a spirally perforated rotating disk that spun in front of an aperture, and the holes admitted light in successive lines. They produced crude pictures made up of 30 lines per frame, compared with 525 lines in contemporary television.

Experiments with mechanical scanning devices were pursued in the 1920s by American inventor Charles Francis Jenkins. He used elaborations of the Nipkow disk to broadcast silhouette pictures from his workshop in Washington, D.C., in 1925. Scottish inventor John Logie Baird followed similar lines of investigation and made a public demonstration of television in 1926, but he also produced only shadow pictures.

Mechanical television by the late 1930s reached the peak of what it could offer. In particular, it did not provide sharpness of detail. Consequently, further advances depended on the development of electronic scanning. Even in the crude stage, electronic systems provided far better transmission, reception, and much clearer pictures.

Electronic Television

Two inventors figured prominently in electronic television developments in the United States: Vladimir Zworkin and Philo T. Farnsworth. Zworkin, who emigrated to the United States from Russia in 1919, developed a crude, but workable, partly electronic TV system in 1923. He evoked the principles of the iconoscope, the basic tube of the television system.

Farnsworth, an Idaho farm boy born in 1906, was 14 before he knew about the existence of electricity, but he grasped its principles rapidly. Within two years, he rejected the idea of mechanical scanning and worked out an electronic television system. In 1930, he developed a new electronic scanning system that made TV pictures suitable for the home.

In 1930, Zworkin became head of a celebrated research group of more than 40 engineers at the RCA laboratories in Camden, New Jersey. Formed from a merger of the television research programs of General Electric and Westinghouse with that of RCA, the Camden team investigated all aspects of electronic television development, solving not only technological problems but also the subjective problem of setting the picture quality standards needed to win public acceptance.

During the 1930s, members of the Camden team tackled and solved the outstanding problems. They progressed to higher line frequencies, increased image size and brightness, and introduced interlace scanning to suppress flicker. They also adapted equipment to use the newly opened VHF band and introduced sets into homes on an experimental basis.

By 1939, the Camden group felt ready for a major public demonstration. RCA chose the 1939 New York World's Fair, with its "World of Tomorrow" theme, as a suitably prestigious and symbolic launching pad for the 441-line RCA television demonstration. Despite the successful demonstration, it wasn't until 1941 that U.S. television systems were approved by the Federal Communications Commission (FCC). Within the year, however, the United States was at war with Japan and Germany. With production of civilian consumer electronics halted, television development had to be shelved for the duration of World War II.

TV Growing Pains

The end of the war in 1945 did not, as some expected, bring an upsurge in television activity, despite a backlog of 158 pending station applications. Investors held back for several reasons. The issue of color television was unresolved, and many experts believed that development should await adoption of a color system. Moreover, potential investors wondered whether the public would buy receivers that cost many times the price of radios. They also questioned whether major advertisers would pay the higher cost of television programming.

However, two favorable developments occurred shortly after the war. The image orthicon camera tube, introduced in 1945, improved camera sensitivity, eliminating the need for the high levels of studio light that the iconoscope had required. The other was that AT&T began to install intercity coaxial cable links, enabling network interconnection. The New York–Washington, D.C., link opened in 1946. Finally, in the summer and fall of 1948, the long-predicted rush into television began.

In 1948, the number of television stations on the air increased from 17 to 48. The number of cities served by television went from eight to 23. Set sales increased more than 500 percent over the 1947 level and by 1951 had surpassed radio set sales. Television's growing pains had not yet ended. The FCC's go-ahead for commercial television had only 12 VHF channels available to serve the entire U.S. As more and more stations began to go on the air, it became obvious that the demand for stations would soon exceed the supply of channels. The FCC had not required enough geographical separation between stations on the same channel to prevent serious co-channel interference.

To forestall a potentially chaotic situation, in September 1948 the FCC abruptly froze the processing of television license applications. However, the freeze did not seriously inhibit television's growth. During the 1948-52 period, the number of sets in use rose from 250,000 to more than 17 million.

When the FCC lifted its freeze in 1952, it allocated channels in the VHF band for more than 500 stations and channels in the UHF band for more than 1,400 stations. Many of the VHF channels were quickly put to use by new stations, but UHF channels were introduced more slowly because receivers and antennas had to be modified to accept UHF signals. UHF television got a lift with the passage of a law requiring all sets manufactured after 1964 be capable of receiving both VHF and UHF signals.

The Advent of Color

After the FCC freeze in 1948, the role of color television and its relationship to the standard black-and-white broadcast remained unresolved. CBS, RCA, and Color Television Inc., each presented a color system for evaluation by the FCC. CBS developed a partially mechanical color television system that was incompatible with the NTSC black-and-white standards. The other two companies claimed their systems were compatible.

In September 1950, the FCC found that the CBS system met its standard for color. As a result, the FCC selected the system for broadcasting, although it kept the door open for the development of a better system. CBS began limited broadcasting in color in June 1951 but seemed to lose interest after November, when the National Production Authority prohibited the manufacture of color TV sets because of a shortage of certain materials essential in the Korean War.

In January 1953, the NTSC adopted specifications for a compatible color TV system supported by most of the major manufacturers. Toward the end of the year, the FCC issued rules for the use of this fully electronic, fully compatible system. It has been in use in the U.S. ever since.

The rival networks eventually tired of the expensive wrangling, and all parties accepted new standards proposed by the NTSC for an electronic system patterned closely on RCA's and thus compatible with sets already in use. This meant that black-and-white receivers already on the market could pick up color signals.

However, color telecasts on a large scale developed slowly because of the cost of receivers and the lack of advertiser interest. Five years after the 1953 FCC adoption of the NTSC standards, only NBC offered programs in color. Full network color production in prime time came only in 1966. By 1972, half the country's homes had color television sets.

Screen Size on the Increase as Home Theater Gains Popularity

When commercial TV was first introduced in the 1950s, a 16-inch set was the biggest available. Twenty years later, the biggest screen was 25 inches. Today, sales of TVs with screens 27 inches or larger are rising, while sale of those with screens smaller than 20 inches are falling. Industry experts cite falling prices, a rebounding economy and the embrace of home theater as the factors for increased sales. In fact, large-screen television, both direct-view and projection, is the primary component of most home theater systems, and the fastest-growing arena of the red-hot color TV business.

New compact home theater systems are attractive to buyers who, in the past, did not have enough space for giant-screen TVs. Even in the larger screen sizes, today's advanced rear-projection TV technology now allows for bigger pictures, while taking less space.

First introduced by Zenith in 1993, the short focal length projection system has allowed designers to replace the boxy look of older projection TVs with high-styled "slim-profile" contemporary cabinets. Examples of this new trend are 60-inch big-screen color TVs that stand less than 28 inches out from the wall. Others include the 52-inch, less than 25 inches in depth, and the 46-inch models, that occupy only 23 inches of space from the wall. These depth dimensions are smaller than most large-screen console TVs.

Projection or large-screen systems come with either a picture tube system or projection tubes. The picture tube system uses a screen-type tube similar to that found in a conventional color TV. This tube has extra brightness that, when focused through at least one lens and reflected through a mirror, will project a picture to a screen. Another format uses projection tubes and is usually of three tubes for the primary red, blue, and green signals which make up a video color signal. The tubes are focused to converge the three colors on a screen, combining to form a complete color picture on the light intensifying surface. The liquid crystal display projection system uses color-emitting semiconductor surfaces that are picked up by high-intensity light sources and projected onto walls or screens.

The Future

There are many devices on the way that will raise the standards for televisions, including high-definition television (HDTV) and flat-panel television screens.

Demonstrated in the United States for the first time in 1981, HDTV promised a picture resolution twice as good as the American standard but at the cost of incompatibility with existing receivers. HDTV research and development in the U.S. began in earnest in 1987, and soon, nearly two dozen different systems were proposed. By mid-1990, however, intense competition and high R&D costs had reduced the field to four major systems: two developed by General Instrument and the Massachusetts Institute of Technology (MIT); one by Zenith and AT&T; and another by a consortium consisting of NBC, the David Sarnoff Research Center, North American Philips, Thomson Consumer Electronics, and Compression Labs. Following the development of a digital system by General Instrument in June 1990, an advisory committee of the FCC expressed its strong preference for a digital standard, which, as a practical matter, forced all the proponents to redirect their R&D efforts away from analog and toward digital HDTV.

On May 24, 1993, the rival groups announced that they had joined forces to develop the next generation of U.S. television technology. The new R&D consortium was dubbed the "Grand Alliance." Members include: AT&T, General Instrument, MIT, Philips, Sarnoff, Thomson, and Zenith. In November 1995, the FCC Advisory Committee on Advanced Television Service (ACATS) urged the FCC to adopt technology developed by the Grand Alliance as the new digital television broadcast standard in the United States.

The broad-based benefits of the Grand Alliance system will bring large-screen, high-resolution displays, and a 19 megabit-per-second pipeline of data (in the same amount of spectrum as today's analog low-resolution TV) into American homes.

Aiming to have quality comparable to HDTV, flat panel television sets feature screens only a few inches thick that can be hung on a wall like a painting. Toshiba, in fact, claims it has a flat-panel TV in its labs with better contrast than images printed on paper. While large displays at first may be targeted toward business and public viewing applications, the real boom for flat panel TVs will be in the widescreen consumer market.

THE HISTORY AND TECHNOLOGY OF THE VCR

HDTV, standard definition television, and the VCR and other recorded media are all inextricably woven together. The existence of television drove the development of the VCR, which, in turn, was one of the developments that drove home theater. Home theater, with its large screens which reveal the defects and shortcomings of standard definition television, is one of the factors that have led home viewers to embrace the concept of HDTV. For those reasons, it is useful for anyone interested in HDTV to have a good background on the development of the VCR. Here is a history of the VCR that was developed by the Consumer Electronic Manufacturers Association (CEMA).

From magnetic disc to today's hi-fi/stereo VCR, the video recording industry has come a long way. Today, millions of home theater owners have the convenience of choosing what they want to watch and deciding when they want to watch it with a hi-fi/stereo VCR or a laserdisc player.

Videodisc and videotape technology goes back to 1927, when John Logie Baird demonstrated what he called "phonovision" using waxed phonograph discs to record TV images.

The next major advance was in 1951, when the first magnetic videotape recorder was demonstrated by Bing Crosby Enterprises. Although it performed poorly, it was predicted that such a recorder would eventually allow full color, stereoscopic pictures and sound to be recorded on one strip of tape. RCA demonstrated a color videotape recorder the following year.

While in the 1950s there may have been dreams of eventually coming up with a videotape system for the home, the market for these first machines was limited to television stations and production houses.

Development in Magnetic Disc Recorders and Home Video Players

Magnetic video recording was responsible for the next innovation in the broadcast market — the creation of "Instant Replay." Its use was largely limited to sports broadcasting because of its expense and the small disc capacity.

In the mid-'60s, a number of manufacturers went after the home videotape industry with no success. Among these were Sony, Ampex, Matsushita (through Panasonic), Philips, General Electric, 3M, and Bell & Howell. Prices of more than $1,000 were the major obstacle, with the units consisting of a combination camera-recorder monitor.

One of the first companies to enter the home video player arena was CBS. In the late 1960s, CBS Laboratories' Dr. Peter Goldmark was responsible for perfecting a device that played film cartridges on the home television set. The system was called electronic video recorder (EVR).

While it was not widely reported, there were difficulties with EVR, including picture weave, vertical and horizontal jitter, and scratched film. In 1971, CBS decided to give up its interest in EVR.

EVR did spark a competitive reaction from CBS's rival RCA. In 1969, RCA demonstrated its holographic color television tape players. RCA was counting on low cost to sell the players, that used inexpensive film for its recordings. However, the picture quality of the device was limited. None the less, consumer appetite for home video recording and viewing had been aroused, and the consumer electronic industry soon responded with new technologies.

Beta vs. VHS

Sony introduced its first Betamax machine for the home market in 1975 for an initial price of $1,300. With this introduction, consumers had the opportunity to record their own programming and move into the world of home movies.

In 1977, Sony's monopoly ended when Japan Victor Company (JVC) introduced the VHS format video cassette recorder (VCR), that permitted consumers to record up to six hours on a single tape. RCA introduced VHS machines and were shortly joined by Panasonic (Matsushita). These machines offered a cassette format that was technically incompatible with Sony's Beta machines.

Gradually, the new VHS machines became dominant in the home VCR market. Longer recording times and increased availability of prerecorded films to consumers stimulated sales, bringing prices down from an average of more than $1,000 in the late 1970s to about a third of that amount a decade later.

By 1978, Magnetic Video was taking steps to license movies on videotape, notably an experimental deal with Fox for some 50 pictures. Today, there are more than 25,000 movies on videotape, according to the Video Software Dealers Association based in Los Angeles. In the mid-1980s, VHS VCRs were available with hi-fi/stereo capabilities that offered Surround Sound encoded audio tracks for use in the home setting.

Introduction of the Videodisc

Magnavox introduced the videodisc in 1978. Although it could be used for playback only, not for home recording, it cost about half of what a VCR then cost and offered a superior picture. Several incompatible formats, using either laser or nonlaser electronic technology, soon came on the market.

RCA bet on the latter approach, introducing its Selectavision in 1981. But mass marketing brought down the prices of VCRs faster than expected, wiping out the cost advantage of videodiscs. Coupled with the inability to record and price competition from VCRs, the videodisc proved fatal in the home market. However, they continued to play a role in training education and business. In 1984, RCA pulled Selectavision out of the home videodisc market.

LASERDISC PLAYERS

While the VCR has achieved commodity status and near saturation, the laserdisc (LD) has had much more modest reach and penetration. Developed by Pioneer in 1978, the first widescale use of the LD was by General Motors for product training of Cadillac salespersons in showrooms. Two years later, Pioneer released the first LD player for home use in the U.S. LD players, which are largely self-contained, include discs that are more durable than videotape. Laserdiscs deliver a 60-percent better picture than conventional VHS VCRs and digital/CD quality sound.

Unlike a CD, a laserdisc holds information on both sides, either half an hour or an hour on each side depending on how the disc was recorded. Most movies are on one disc divided into two sides. The difference between film on laserdisc and on videotape is more than the physical medium: a laserdisc is capable of 425 lines of resolution, the highest resolution offered in any National Television Standards Committee format, better than broadcast TV and almost twice as good as the VHS videotape.

Another difference between movies as presented on laserdisc versus videotape is the availability of special editions, such as discs that have extras including original movie trailers, documentaries, and other behind-the-scenes information. Many directors reedit their films, add footage, and produce versions available only as director's editions on laserdiscs. Many discs have separate audio tracks that include commentary by a film scholar or by the director while the movie plays.

There are nearly 10,000 movies available on laserdisc and almost every major Hollywood film is released on laserdisc as soon as it is released on videotape, according to *Laserdisc Newsletter*, published by Douglas Pratt Publishing in Long Island, New York.

ENJOY GREAT PICTURE AND SOUND ON HOME THEATER

The benefits of stereo soundtrack that today's hi-fi/stereo VCRs and laserdisc players offer are best enjoyed in a home theater. Sales gains in hi-fi/stereo VCR show home theater to be one of the hottest consumer electronic products of recent years.

1997 AND BEYOND

by Gary Shapiro
President
Consumer Electronics Manufacturers Association

The Consumer Electronics Manufacturers Association has a unique perspective on the history and development of consumer electronics, and a clear vision of where this exciting segment of our lives is headed. This presentation by Gary Shapiro, president of CEMA presents his vision of the future of consumer electronics.

Historically speaking, few industries can claim the same kind of cultural and economic impact that defines the consumer electronics (CE) industry. Throughout the past half-century, and particularly the past two decades, the consumer electronics business has transformed society radically, making accessible previously unimaginable worlds of information, entertainment, and personal empowerment to millions worldwide. It is no exaggeration to say that the latter half of the 20th century can be characterized as the age of electronics.

But for all of our industry's impact, the future looms more monumentally than the present foundation of established successes. Existing technologies are becoming more powerful and available to a wider range of consumers. Emerging technologies create possibilities that will result in new product categories, and along with them, plentiful business opportunities.

The 21st century will be known as the digital age, and the cornerstones for it are being laid right now in the consumer electronics industry.

Perhaps the only way to predict the effects of this technological tidal wave would be with a look at the past, present, and future of our industry's accomplishments. 1997 was a watershed year for the CE business. It pointed to a pattern of growth and ebb in several key sectors, revealing the evolving state of consumer sophistication and overall economic trends. The industry experienced yet another year of outstanding growth. Total sales topped $72 billion, a 4 percent increase over 1996. This increase, while not surprising in view of the industry's exceptional growth patterns during the past five years, is remarkable considering consumer consternation (and confusion) regarding new technologies due in the market beginning this year.

Some of these technologies are radical recastings of popular extant product categories. The category of DTV (digital television) and its highest-level iteration, high-definition television (HDTV), is perhaps the best example of the evolutionary fork in the road that both manufacturers and consumers face in coming years. Television sales, both for direct-view and big-screen projection sets, plateaued in 1997, bringing the overall category of video products to a final sales tally of more than $14 billion, down from more than $15 billion in the previous year. But this figure is misleading on two counts, and both can be traced directly to the impending DTV innovation.

Consumer awareness of DTV and HDTV is high, considering that the technology has not yet officially launched. This has led not only to some consumer hesitancy toward the popular TV category, but correspondingly, has led to significant retail price breaks for existing analog TV products. This represents a boon to savvy consumers willing to invest in today's TVs, which are at the highest level of advancement in history, and a tremendous "bang for the buck." CEMA and the industry have worked hard to educate consumers about the value and utility of today's analog TV sets, which will represent the lion's share of video entertainment for the next decade.

1998 marked the long-awaited launch of a new format, the Digital Versatile Disc, or DVD, sometimes referred to as "Digital Video Disc." Envisioned as the next-generation carrier for all manner of digital entertainment, including movies, music, and videogames, DVD is the result and symbol of a new level of "converged" interests between the veteran consumer electronics companies and the computer industry. More than a year of technological debate surrounded the introduction of DVD, with the final outcome a "cross-platform" digital carrier that can work both in traditional CE settings, like a home theater, or on the desktop in a PC environment.

Despite the fits and starts of the DVD launch, consumer acceptance of the new format in 1997 was triumphantly high, representing sales of $171 million in less than a year. DVD, now firmly established in consumers' minds as a legitimate new format, experienced better sales performance in its first year than either color TV or the CD format, both perennial stalwarts of the CE industry.

The home audio category also was subject to the "wait-and-see" attitude consumers had toward evolutionary new products in the video category. Technological advances in digital consumer equipment perhaps are most mature in the audio field, where the compact disc—the first carrier of the digital age—celebrated its 15th anniversary in 1998. The CD format, so well established in consumers' lives, now is being updated with a series of hardware and software that points to an even further extension of this category in everyday life. On the software side, the industry is at work developing a new, high-resolution audio disc that will utilize DVD drives as the carrier for next-generation digital multichannel sound. On the hardware side, consumers have shown remarkable receptivity to virtually every subdivision of the present CD category—portable players, CD changers, digital-to-analog converters for the specialty audio market, and even recordable CD technology. This latter innovation is one that consumers have been awaiting anxiously, and one that undoubtedly will serve to further the sales gains made in the CD category, which in itself experienced a more than 10 percent increase in unit sales in 1997 compared to 1996.

While home entertainment products experienced a year of uncertainty, there was none regarding the other side of the "convergence" spectrum—the market for information and communication products, which totaled more than $31 billion in sales, up from $28 billion the year before. The small office/home office (SoHo) trend no longer is regarded as an alternative to the traditional workplace. In many cases, corporations now offer "virtual commuting" as an enticement to employees, and as a way to retain a skilled workforce in the face of shifting economic patterns. Home office workers now are a significant part of the U.S. workforce, and the products that power these SoHo's, both at the desk and on the road, have exhibited substantial market penetration. For example, 42 percent of U.S. households have at least one personal computer. These are joined by a 24 percent penetration of modems for online work, and an additional 10 percent of households with a dedicated fax machine. To enable this work and still retain the ability to juggle everyday household chores, Americans increasingly are turning to cordless phones, which now boast 70 percent penetration, and experienced a stellar 1997 with an astounding 42 percent dollar sales increase over the previous year.

This portability naturally extends into the mobile electronics sector, which now is becoming as all-pervasive as home electronics. Technological evolutions and increasingly affordable price points have put the tools of mobile business into the hands of millions of Americans. Pagers are a perfect example, with the category showing a better than 25 percent increase in sales in 1997 compared to the previous year. Wireless (cellular and PCS) telephony is a big star, too, with factory sales of $2.7 billion for 1997. As Americans increasingly incorporate the tools of business into their cars and briefcases, they continue to bring the tools of entertainment into the same contexts. While aftermarket autosound sales were nearly identical in 1996 and 1997, sales of factory-installed autosound increased by almost 10 percent, to slightly more than $2.7 billion in sales. These figures not only point to a robust economy where new auto sales are on an upswing, but to the growing demand for high-quality consumer electronics in the automobile cockpit, which increasingly is regarded as an extension of the home and office.

While only a few years ago multimedia was something that required specialized equipment and knowledge, today the medium is well known and understood by the American consumer. The personal computer has been a leader here, with more than 26 percent of households equipped with a CD-ROM drive. Perhaps the most stunning gains in multimedia were made in the area of digital photography, a relatively new technology to the consumer market. Growth here has been explosive, with the 1997 sales figure of $392 million more than doubling the $177 million of 1996. But even as the market for evolving technologies explodes, more established multimedia categories, particularly dedicated videogame consoles, still are vital and growing. Despite inroads from new media like DVD and online gaming on the Internet, existing videogame hardware and software experienced an almost 9 percent increase over the previous year's sales figures.

Perhaps the most interesting trend through all this activity will be the impact of the continued convergence of the major consumer electronics categories. This trend, which merges the interactive capabilities of computer products with the functionality and everyday utility of traditional consumer electronics products, points to outstanding business opportunities for both existing and yet-to-be-invented categories. In the former camp, there will be evolutions of existing product architectures that have been retrofitted and extended with interactive capabilities. The changes will come in both the entertainment and information arenas, and already are exemplified by Internet-accessible set-top boxes that turn a TV into a Web-surfing machine; and by personal digital assistants that can connect to public or private networks through wired or wireless modem connections.

The latter camp of products, yet to be devised, represents almost limitless potential for the CE industry. Already there are "set-top boxes" in development that transcend today's slow copper phone lines with broadband access through cable TV lines and direct broadcast satellite (DBS) dishes. These devices will transform television from a passive to an active medium, not only providing entertainment and information on demand, but creating new media where the audience is an active participant in selecting, customizing, and perhaps even creating the broadcast content. As the most common denominator between the two camps of the convergence spectrum, video pictures will play a large part in these coming technologies. The combination of high-speed, broadband access and effective data compression techniques like MPEG will result in new roles for video. Wired video telephony already is a reality; with the usual curve of technological advances and corresponding price drops, it isn't hard to imagine such items as PDAs, cellular products, and perhaps even watches that feature video as a selling point.

The convergence of technology in the consumer market will mean a convergence of interest between two pillars of the American economy—the PC and CE industries. While both sides often ally toward their own methods and strategies, it is clear that their interests—and with them, their products—will continue to intertwine. In the coming years, the consumer will expect standards and protocols that will allow information and entertainment to flow from product to product, in the home, at the office, in the car, in a briefcase, or in a backpack. Software will be devised to work as part of a fluid environment, independent of hardware carrier. Whether standards emerge from today's existing software operating systems, or from a new one yet to be released or even developed, is one of the key questions of the digital age. As the track is laid for the infrastructure of our digital times, consumers will expect the same kind of functionality, ease of operation, and reliability they have come to expect from their current consumer electronics products. Meeting this challenge is the common goal of both the PC and CE industries.

As we speed along to this century's conclusion, the consumer electronics industry will continue its influence as the defining signifier of our age. It remains a strong and growing entity, informed not only by continuing technological advances, but by a record of consumer satisfaction and trust that is the envy of other industries. In these times, it's hard not to be optimistic about the CE industry, and we all look forward to its continued role as a shaper of both today and tomorrow.

6

TECHNICAL QUESTIONS
AND ANSWERS FOR DIGITAL TELEVISION

When a technology is as new as high-definition television is today, sometimes it's difficult to even be able to learn enough to be able to formulate questions, answers to which can help clarify the concepts. These questions and answers were developed by the Consumer Electronics Manufacturers Association to help people understand the concepts on which HDTV is based.

Q: What is the agreed upon definition for high-definition TV (HDTV)?

A: There is a generally agreed upon definition of HDTV, i.e., approximately twice the vertical and twice the horizontal resolution of today's TV. Because HDTV also has a wider screen, HDTV pictures contain about five times as much picture information (picture elements or "pixels") as conventional TV—twice as much vertically, twice as much horizontally, and then a bit more to fill the wider screen. HDTV usually also includes multiple channels of digital surround sound as well, in the case of the Advanced Television Systems Committee (ATSC) DTV Standard based on the Grand Alliance HDTV system, the standard offers 5.1 independent channels of CD-quality surround sound. More information is available at http://www.atsc.org.

To many people, high definition means 1,000 vertical scan lines in the picture, but things aren't quite that simple. For the same number of lines, a progressively scanned picture offers a higher effective resolution than a picture using interlaced scanning where only half of the lines are shown at a time. Thus, there are two fundamental HDTV formats in the ATSC DTV Standard—1080 lines x 1920 pixels per line; and 720 x 1280 pixels per line. These two fundamental pixel arrays are used at three different frame rates: 60 frames per second for live video, and 24 and 30 frames per second for material originally produced in film. Thus, there are six HDTV formats in the standard. All of them use progressive scan except one—the 1080 x 1920 at 60 frames per second. This particular format requires the most bits, and the only way we can fit this format into a 6 MHz terrestrial broadcast channel is to use interlaced scanning, which is essentially a further compression technique which reduces the bit rate by only reproducing half of the picture at a time.

Summary of HDTV formats:

1080 x 1920 ..... 24P, 30P, 60I (frame rates)

720 x 1280 ..... 24P, 30P, 60P (frame rates)

Although knowledgeable people often refer to the 1080 interlaced format as providing the highest level of resolution, in fact, it's more complicated than that, 720 progressive is roughly comparable in effective vertical resolution to 1080 interlace, although the 1080 x 1920 format will offer greater horizontal resolution. And, of course, the 1080 x 1920 progressive formats will provide the highest spatial resolution, but at lower frame rates, i.e., at lower temporal resolution. (Everything about digital TV is a trade-off between various factors. Here, you can get higher spatial resolution at the expense of lower temporal resolution. Different types of programming would cause you to choose different trade-offs, thus the inclusion of multiple formats to choose from in the Standard.)

Q: What is the difference between digital television (DTV) and high-definition TV (HDTV)? Can the terms be used interchangeably?

A: Digital Television (DTV) is a broad term encompassing all types of digital transmission. HDTV is a subset of DTV indicating 1000+ lines of horizontal resolution. Another type of DTV is Standard Definition Television (SDTV) which has picture quality equivalent to (or slightly better than) a good analog picture. Since it requires less data to generate an SDTV picture, a broadcaster will be able to transmit multiple channels of SDTV within its 6 MHz digital channel.

What broadcasters can send in a 6 MHz channel is frequently misunderstood and misstated. It depends entirely on the type of program material that is being sent. For the toughest live video, i.e., fast-action sports like basketball, it does require most of the channel to send a single HDTV program. However, for more typical video like game shows, you could usually send one HDTV and one SDTV program concurrently over the channel.

For film-based material (all movies and 70 to 80 percent of prime-time programming), you can send two HDTV programs simultaneously.

Here's a summary of what you can send concurrently over the channel:

Sports/Action Video .... 1 HD, or 2-3 SD

Typical Video .... 1 HD plus 1 SD, or 4 SD

Movies/Slow video .... 2 HD, or 6 SD

In addition to all of this, you can carry immense amounts of data over the channel, either by dedicating some portion of the channel capacity, or by using the channel opportunistically. The Grand Alliance did an experiment with a BMW commercial produced in HDTV. When the car swooshes across the countryside, it takes most of the channel to reproduce top-quality HDTV pictures, but when the BMW logo is shown for a few seconds, much of the channel capacity is available for data. Over the course of this 51-second commercial, 60 Mbytes of data (40 floppy diskettes worth!) were transferred through the channel simultaneously with the delivery of a high-quality HDTV commercial. Thus, the challenge for using the DTV transmission capability for information services is not in squeezing data through, but it is a marketing challenge to figure out which data applications consumers will desire.

Q: Does high definition only mean interlaced, or could it also mean 1080 lines of progressive scan?

A: The term "high definition" could apply to 1080 lines of progressive scan. While the technology does not yet exist to allow broadcasters to transmit 1080 lines of progressive scan in the digital channel, such technology should be available in the intermediate future.

High definition most definitely does not only mean interlaced. Indeed, five of the six ATSC HDTV formats are progressive. One of the most exciting things about the Standard is that film-based material, including all movies and 70 to 80 percent of prime-time programming (like *ER*, etc.) can be shown in 1080 x 1920 progressive scan (at 24 frames per second or "Hz"). And by the way, all film material will be automatically transmitted using progressive scan formats. It would be inefficient and more difficult to use interlaced for film-based material. Thus, from day one, most digital TV programming will be transmitted in progressive scan.

As noted, we cannot yet fit 1080 x 1920 progressive at 60 Hz into a 6 MHz channel, and it's not even close right now. We have just started work within ATSC on a layered coding approach to enable us to do this in a backward compatible way, but additional invention is required. It's hard to say when this might be possible, but it's very unlikely in the immediate future.

Q: What are the intentions of broadcasters to use both progressive scan and interlaced formats?

A: It's a complicated issue for broadcasters to decide which formats to use, but it's virtually certain that they will use a variety of formats for different purposes. CBS has spoken out clearly in favor of 1080, interlace for video, progressive for film. NBC has stated clearly that it will offer substantial amounts of HDTV, but has not clearly indicated whether it will be 1080, 720, or a mix of both. ABC has said it will offer HDTV, and has shown a preference for progressive scan, i.e., 720 progressive, but that could change. Fox originally spoke in favor of SDTV to the exclusion of HDTV, but last year Rupert Murdoch seemed to indicate that a quick look at the business case for multiple SDTV had rekindled Fox's interest in HDTV. There is no pivotal cost issue on progressive vs. interlace. You will see both progressive and interlace in both HDTV and SDTV.

FREQUENTLY ASKED QUESTIONS ABOUT DIGITAL TELEVISION AND CABLE MUST-CARRY OBLIGATIONS

At the moment, HDTV is being received over the air, via broadcast receiving antennas. Cable televisions are not, at the moment, involved in carrying HDTV signals. Following are the views of the Consumer Electronics Manufacturers Association on this important aspect of TV signal delivery, in Q and A form.

Q: Why is it important that cable companies carry broadcasters' digital television (DTV) signals?

A: It is important because more than two-thirds of American homes currently receive their living room television programming though a cable system. Local monopoly cable companies must not be allowed to exploit this "gatekeeper" position by degrading broadcasters' HDTV signals, or by not carrying digital signals at all.

Cable refusal to transmit broadcasters' full digital signals would severely imperil the transition to digital broadcasting. Broadcasters will be highly reluctant to begin digital broadcasting if they are unsure the broadcasts will reach two-thirds of their viewing audience. Consumers will not be eager to buy a first-generation digital receiver if they are unsure whether it will work with their cable system. Only by requiring cable companies to carry broadcasters' full digital signals can we ensure a successful transition to DTV, and a rapid return of the analog spectrum to the government.

Q: Does the law require cable companies to carry broadcaster DTV signals?

A: CEMA believes that Section 614 of the Communications Act requires that cable companies carry the signals of local broadcast television stations.

This means that cable companies are required to provide their viewers with the entire 6 Mhz digital channel used by broadcasters, delivering all of the broadcaster's digital bits intact.

Q: Are cable companies prohibited from taking a broadcaster's HDTV signal and "downconverting" it to a lower resolution format?

A: Yes, because this would be a "material degradation" of the broadcaster's signal prohibited under the Communications Act.

American HDTV owners would be severely penalized if cable companies were allowed to take a broadcaster's high-resolution signal and convert it to a format only slightly better than current analog TV. This would be the same as (if not worse than) a cable company's taking a network broadcast like the Super Bowl and delivering it to all its viewers with a black-and-white picture and mono sound.

Q: If a broadcaster multiplexes its channels, should the cable company be required to show all of them?

A: Yes. Broadcasters will be offering a variety of different DTV services including HDTV and SDTV "multicasting" at different times of the day.

It is critical that local cable operators not use their monopoly position to act as gatekeeper, determining what DTV broadcasts television viewers are allowed to see by allowing access to selected DTV broadcast transmissions and denying access to others.

Q: Do cable operators have the capacity to carry broadcasters' digital signals?

A: Yes. Cable capacity has grown significantly over the last few years. Already, two cable MSOs—Cablevision and Time Warner Cable—have announced plans to provide HDTV programming to subscribers.

NEW CHOICES FOR LINKING CABLE AND TV SETS

Background

The Consumer Electronics Manufacturers Association (CEMA) is proceeding with four technical solutions to link cable and other set-top boxes to television receivers: a 1394 interface, the RF remodulator interface, a component video interface, and the National Renewable Security Standard (NRSS) interface. These efforts are designed to ensure that consumers who buy a DTV set in the future will be able to receive digital cable and use their DTV sets with other digital technologies like direct broadcast satellite (DBS), digital versatile disc (DVD) players, digital VCRs, and computers.

The four interfaces will provide consumers and manufacturers a myriad of choices for connecting their TV sets to cable systems. With four interfaces, consumers will be able to choose from simple, low-cost connections to high-end, feature-rich models.

How do the Interfaces Differ?

1394 Interface:

- The Institute of Electrical and Electronics Engineers (IEEE) and International Electrotechnical Commission (IEC) have created international 1394 standards: IEEE 1394-1995 and IEC 61883. Many other organizations are near completion on other related standards, and these standards are now far enough along that companies are creating "suites" of protocol to implement these standards and specifications. CEMA's R4.8 1394 Interface Subcommittee is working rapidly to meet Federal Communications Commission (FCC) Chairman Kennard's goal of an industry specification by November 1, 1998, for implementation of 1394 to connect cable set-top boxes and DTV.

- The 1394 interface creates a two-way digital bus architecture.

- Digital devices can be daisy-chained together as opposed to using point-to-point connections. Unlike the other three interfaces, 1394 is not only a cable set-top box to receiver connection; it can connect several digital devices together such as digital versatile disc (DVD) players, digital VHS players, Dolby Digital surround sound processors/receivers, digital converter boxes and other digital consumer electronics products.

- CEMA is investigating copy protection methods and their impact on receiver designs.

- Of the four interfaces being developed, 1394 is the most high-end, feature-rich interface.

RF Remodulator Interface (EIA-762):

- This standard has been published as EIA-762.

- The RF remodulator interface links one digital device (such as a digital cable set-top box) to a digital TV receiver.

- Using a one-way, point-to-point digital connection, the interface can translate the cable industry's transmission language, Quadrature Amplitude Modulation (QAM), into vestigial sideband (VSB), the transmission standard accepted by the Advanced Television Systems Center (ATSC) for use with DTV.

- The interface can support one-way digital copy protection schemes.

Component Video Interface (EIA-770):

- This standard has been published as a three-part document, EIA-770.

- The component video interface can link a cable set-top box to a TV receiver.

- The interface uses a one-way, point-to-point analog connection.

- The interface can support analog copy protection technology.

National Renewable Security Standard (NRSS) Interface (EIA-679):

- This standard will be published as EIA-679.

- The NRSS interface applies to digital cable-ready TV receivers.

- In the NRSS interface model, a smart card or PCMCIA card provides access security to pay and subscription cable or satellite TV services, telephony and all forms of electronic commerce.

- The interface can support two-way copy protection schemes.

FREQUENTLY ASKED QUESTIONS ABOUT DIGITAL TV

Q: What is the difference between Digital TV, HDTV, and SDTV?

A: Digital TV (DTV) is the umbrella term used to describe the new digital television system adopted by the FCC in December 1996. DTV in its broadest terms includes High-Definition Television (HDTV), Standard Definition Television (SDTV), and a host of other potential applications. HDTV defines certain minimum performance attributes that deliver approximately twice the resolution of current TVs. SDTV refers to a system that provides a display resolution lower than that of HDTV.

Q: Do consumers want HDTV?

A: Absolutely. From black-and-white TV to color, from phonographs to CDs, from videotapes to DVD, the American consumer has never passed up the opportunity to take advantage of improved video and audio performance. HDTV is no different. Surveys of consumers who have seen the technology indicate an overwhelming interest in its revolutionary picture and sound quality improvements and its potential for interactivity (see CEMA/Verity HDTV consumer focus group research).

Q: Why would I buy a regular analog TV set with DTV coming?

A: These sets will work for decades. For broadcast signals they will work until at least 2006 and then low-cost set-top converter boxes will be available. Even if the broadcast analog signal is ended in 2006, a $1,000 set bought in 1997 would cost about three cents a minute of average use for the next 10 years.

Today's TV sets are a tremendous bargain. Large-screen prices are the lowest in history. Prices have dropped an average of $100 per year.

Q: When the analog signal shuts off, will people throw out obsolete sets?

A: Sets won't be obsolete. First, the TV sets will work for decades with cable and home satellite, and forever with current VCRs, DVD, and other products. Second, for broadcast television, we expect low cost (under $150) converters to be available when digital signals replace analog signals.

As soon as the transition to digital begins, current sets can receive both the analog and digital broadcasts with the addition of a set-top box.

Q: What does the 2006 termination really mean?

A: 2006 is a "target." In Washington, Democrats and Republicans want to balance the federal budget by the year 2004. Auctioning the analog spectrum has become a mantra for politicians as one method to help offset a portion of the budget deficits. In reality, few people really believe that DTV penetration by 2005 will allow the shutting down of the analog signals by 2006. Federal legislation approved in 1997 allows for a continuation of the analog signal in markets where penetration of digital sets is less than 85 percent.

Even though the budget is a politically explosive issue, broadcasters are committed to making huge investments to protect their existing franchises.

Q: How many television sets are there in use today?

A: There are 250 million TV sets in use in the United States today.

Q: How soon will consumers be buying DTV?

A: Conservatively, about 30 percent of U.S. households are projected to own a digital set by the year 2006.

Q: How are TV screen sizes measured on HDTV?

A: As with today's analog sets, HDTV screen sizes are measured diagonally. But, because the aspect ratio on HDTV is 16:9 instead of 4:3, like analog sets, an HDTV set with a comparable height measurement to an analog set will have a larger screen. For example, today's 35-inch sets have a 28" X 21" screen measurement. An HDTV set with an equal screen height of 21 inches will have a 42-inch screen size measurement.

Q: What programming will be available when HDTV hits the retail market?

A: Three of the four major broadcast networks committed to at least some HDTV transmissions this fall. PBS will also broadcast programming in HDTV, as are HBO and Discovery Channel. Fox is planning to use a standard-definition format for all its digital TV transmissions, but may be considering high-definition formats for some of its prime-time lineup. DirecTV, a digital satellite provider, plans to provide two channels of HDTV in the fall.

FREQUENTLY ASKED QUESTIONS ABOUT DIGITAL TELEVISION AND OVER-THE-AIR RECEPTION

Digital Television (DTV) will be available to consumers by over-the-air broadcast beginning in November. Stations in 22 markets across the United States are expected to begin over-the-air digital broadcasts in 1998.

All commercial stations in the nation's top 10 markets were scheduled to be on the air by May 1999, and all stations in the top 30 markets by November 1999. All other commercial stations must be broadcasting DTV by May 2002.

Q: Will DTV be receivable by antenna?

A: Most homeowners will be able to receive broadcast DTV through an antenna. However, as with analog television, the over-the-air signal may be subject to terrain interference in certain areas.

Q: Will consumers need new antennas to receive digital television?

A: No. The same antennas in use today to receive analog transmissions will also work with DTV; however, the nature of the DTV signal makes it important to have the correct type of antenna installed.

Q: Will indoor antennas work with DTV?

A: Yes. However, as with analog television, outdoor antennas will generally be more effective than indoor antennas.

Q: Can local governments or homeowners associations prohibit homeowners from putting up outdoor antennas to receive DTV?

A: No. The FCC recently barred unnecessary local government or homeowner restrictions on television antenna placement. These new rules will make it much easier for homeowners to put up exterior antennas to receive digital broadcast services.

Q: Is anything being done to ensure that consumers will have the highest quality reception of over-the-air digital television?

A: CEMA and the National Association of Broadcasters are working together to identify the factors necessary for quality reception of over-the-air DTV. The resulting guidelines will help retailers, transmitter manufacturers, broadcasters, television manufacturers, and antenna manufacturers ensure consumers the highest quality reception possible.

ANTENNA SUBDIVISION

CEMA's new Antenna Subdivision is working with retailers to ensure that consumers are provided with the most effective antenna for where they live. More information is available at http://www.cemacity.org.

Q: What can consumers do to increase antenna performance?

A: Many consumers can increase performance by merely replacing their current old or damaged antenna. Today's antennas come in many sleek and discrete designs and can provide remarkable reception.

Q: Can DTV be received by any means other than over-the-air broadcasts?

A: Yes. Some digital satellite systems have already committed to providing DTV and HDTV. In addition, while cable is still formulating its DTV plans, providers such as HBO and the Discovery Channel have announced that they will be delivering DTV and HDTV programming.

7
HOME THEATER BY DESIGN

Because HDTV is so similar in concept to today's moving picture theater systems—wide-screen format, high-definition picture, high-fidelity surround sound—most viewers who purchase HDTV receivers will probably want to set up their sets in some kind of home theater arrangement to take advantage of all of those benefits. With that in mind, the following is a generalized set of guidelines developed by the Consumer Electronics Manufacturers Association for readers who may want to set up their own HDTV home theater systems.

As home theater continues to increase in popularity, many consumers—whether it's at the start of the football season or for a love of movies—will face the inevitable urge to invest in a home theater system. But where do you begin, and how do you find a system that's right for your home?

High quality home theater systems are available in all price ranges, beginning at around $1,500, up to six-figure range. You can also find home theater systems that will mix well with just about any decor. In fact, by working with a custom installer and an interior designer, you'll see that you don't have to pay outrageous prices to get great quality, and you don't have to compromise on aesthetics. Here are some basic tips to get you started.

Begin by taking a look at what you have now—a TV, hi-fi VCR, laserdisc player, loudspeakers, A/V receiver. Determine the weakest link in your system, and start upgrading from that point. A custom installer can help you here, and may be the best person to tell you which product you should replace first and what products will blend well with the remaining ones.

Then, consider where the system will be located and the parameters of the space. Do you want your home theater in the living room, or will you create a room exclusively for the system? What size television set will you need? Try to find something that fits the room. You don't want to put an 80-inch television set in a 12x12 room. Normally, the viewing distance should be three to four times the screen size.

Another important, but often overlooked, element of home theater is lighting. "Whichever room you select, make sure the lighting is right for the equipment—and that the equipment is right for the room," explains Wayne Puntel, president of AudioCraft, an Ohio-based installation company. "For example, if you have a front-projection television set in a well-lit room, you won't be able to see the screen." If you decide on a dark home theater media room, you might think about using a projector-type set, which is readily adaptable for such an environment.

Next, you need to get the right sound. Quite often, people hear a demo in the store, but when they get the equipment home, the sound is different. It's important to pay close attention to the room because what sounds crisp and clear in one room may not sound as well in others. While sound is subjective to each person, each room also has a different acoustical personality. Puntel adds, "It's imperative that the equipment is placed properly to maximize the effect. If you compromise the placement of your equipment, you will compromise the quality of your system."

When planning the layout of a home theater system, you'll want to consider the type of floor covering and the shape and size of the room. Once you've decided on a system, an interior designer can work with the installer to determine which space to use, how much space to use, and whether or not to place the projector inside your coffee table! Do you want plush velvet seats? Or how about gargoyles to create an authentic theater effect? A designer can put the finishing touches on your home theater system and camouflage any components that don't blend in with the decor. If you really want to have some fun, you might even ask your designer about incorporating a popcorn machine.

A main advantage to working with a designer and an installer, and to owning a home theater system made up of several components, is that you can improve a single element of the system at a time. As newer technologies emerge, such as Direct Broadcast Systems (DBS) and Digital Versatile Discs (DVD), the system can grow and mature with you over many years. This makes a high-end home theater system a great value in addition to a superb source of musical and video entertainment.

Do-it-Yourself Home Theater Installation

According to the Consumer Electronics Manufacturers Association (CEMA), more than 80 percent of the country's home theater owners installed their home theaters themselves.

All it takes is a little planning, a few hand tools, and a basic understanding of the equipment that goes into the home theater system. Here's how to do it:

The Room

Begin by selecting the room for your home theater. You'll want a room that is large enough to accommodate the equipment and the seating area. Ideally it will be 12 feet by 14 feet or larger and have one wall that is free of doors or windows. If the room has a focal point such as a fireplace, begin planning your layout around that. Otherwise, you will make your home entertainment system the focal point.

Utilities

Your system will require adequate electric power with outlets on the wall behind where your equipment will be. The circuit should not be used for major appliances, air conditioners, or the furnace, and should be properly grounded. If you have cable service or an antenna, you will want to ensure that the cable or antenna outlet is on the equipment wall or is near enough that the cable can easily reach the system. Similarly, if you have a satellite service, it will be necessary to be able to run a signal feed line from the satellite dish to the system.

The Equipment

A home theater typically includes a television with a screen 25 inches or larger, a hi-fi stereo VCR, an audio-video receiver, and at least four speakers. More elaborate systems include laserdisc players, a center channel speaker, and a powered subwoofer. Deciding where to place this equipment will be one of your most important decisions.

The Arrangement

The primary consideration will be the relationship between the television screen and the seating area. Ideally, the primary seating unit (a couch for example) will be parallel to the television screen and face directly at the screen (the center of the couch should be on line with the center of the screen). The distance between the television and the couch depends on the size of the screen. As a general rule, the distance between the viewer and the television should be about four times the size of the screen. For example, if your screen is 35 inches, the seating area should be 10 to 11 feet away.

Speakers

Proper placement of the speakers is critical to your enjoyment of the surround sound capabilities of your home theater. You will have at least four speakers and may have six or more. Here's how they should be arranged:

• The front speakers should be on the same wall as the television, facing the seating area. Ideally, the speakers will be about two feet off the floor, and the span between them should be the same as the distance from the television to the seating area.

- The rear speakers should be placed on the wall behind the seating area, or on the side walls in line with or behind the seating area, high enough from the floor to have an unobstructed path to the listener.

- The center channel speaker, if you have one, should be above or below the television. Many companies manufacture speakers specifically for center channel use. These speakers are magnetically shielded so as not to interfere with the television picture color. If you are using a speaker that was not designed specifically to be a center channel, do not place it on the television since the magnets in the speaker can cause the TV picture to become green and purple.

- A subwoofer, which produces the thundering bass that is an important part of many movie soundtracks, should sit on the floor, near a corner if possible.

Other Components

The location of the other components in your home theater system is not critical to the quality of the experience. They should be convenient for you to reach, and you will save money on cable if you place them near the television. Obviously, they will need to be near proper AC outlets.

Cables

A common mistake many new home theater owners make is selecting inadequate cables for their system. Using standard electrical wire to connect the speakers, for example, will result in audio quality that is probably significantly less than your system is capable of delivering. The key is not to cut costs on cables. Some installers recommend spending 10 to 15 percent of your system budget on wiring. Your cost will depend on several factors, including the distance between the components and the speakers. Buy your speaker wires and cables from a reputable audio-video specialty store. Be ready to tell the salesperson what kind of equipment you have and how far each cable will need to run, since greater distances require heavier cable. The cables that run between the audio and video components could cost between $35 and $50 each. Good quality speaker wire starts at $1.75 per foot, and you could spend as much as $7 or more per foot.

Hiding the Cables

Probably the most vexing aspect of the home theater installation is figuring out what to do with the cables. It is frequently not a major problem connecting the television and audio and video components since they are often in the same cabinet or close to each other on a common wall. The real puzzler is the speaker wire, especially the cables running to the rear speakers. Some home theater owners have solved the problem by purchasing remote wireless speakers that don't require any cable. Others run the cable through an attic or unfinished basement. And some have carefully removed the room's baseboards, run the cables along the studs at the bottom of the drywall and then replaced the baseboards.

However, the two most common methods are running the cables, secured with insulated "U" shaped electrical staples, along the baseboard. Or putting the cables under a carpet. Either method is adequate. Wires along the baseboard can be painted and will not be noticeable. And flatline speaker cable, which is thin and flat, will not cause any carpet lumps or ridges. A note of caution: While it is acceptable to place speaker wire under a carpet, you should never do that with electrical wire since foot traffic on the carpet can wear away the insulation and cause a short and possibly a serious fire.

Cabinets

Until just recently, you had few options for cabinets for your home theater. But today, many furniture manufacturers are offering full cabinets, which include space for large-screen television sets, in styles to fit virtually every decorating scheme. In some cases, furniture makers and consumer electronics companies have joined forces so you can buy your equipment already installed in a handsome cabinet. Or, you may want to visit a furniture retailer and select a cabinet that is just right for your home decor.

Lighting

Once you have completed the installation of your system, you will want to be sure you show your home theater in the best possible light. For maximum enjoyment, the room should be dim but not dark. Make sure that table lamps and ceiling fixtures do not reflect in the television screen, that drapes or blinds can block sunlight from coming in through windows, and that there is a soft light on the wall behind the television set.

Then, dim the lights, make the popcorn, and enjoy the movie theater experience without leaving your home.

The Life Cycle of a Motion Picture

And while we're talking about home theater, let's have a look at the life cycle of a motion picture, as seen through the eyes of the Consumer Electronics Manufacturers Association.

Although home viewing is the most popular way to see movies, most motion pictures still start out as they always have—in theaters. The movie's life begins with opening weekend, where its seen at theaters in major cities, and can end up several months later re-created in millions of home theaters.

A motion picture's value to its owner changes as the amount of public access to it increases and as it ages. A successful movie can remain in theatrical release for six months or more, and a failure can be gone from theater scenes after two weekends. But movies can have a rejuvenated value each time they are sold to a delivery system or format.

The life cycle of a motion picture in the United States consists of tiers. Theatrical exhibition is the first tier, and marketing to subsequent formats is often determined by the movie's commercial success in the theater. The first weekends of a theatrical run often begin in New York, Los Angeles, and Toronto.

After three or four months in the theaters, the movie is released to airlines and hotels. The second tier in a picture's life, perhaps six months after theatrical release, is when it is available in video stores as an over-the-counter videocassette sale or rental item and has virtually disappeared from neighborhood theaters.

Within 30 to 60 days of home video release is home pay-per-view. Next is free cable, which is released six to nine months after home video.

More time needs to pass before the picture can be shown on television, either local or network. This is known as the fourth tier.

Pictures originating as made-for-network such as broadcast, cable, or pay television movies are subsequently sold via domestic television syndication as a further product for local TV station libraries and are sometimes released theatrically in certain other countries.

There are, of course, some pictures that alter this pattern of release depending on what the market will bear. In the U.S., *Rocky Horror Picture Show* pioneered the concept of weekend midnight showing and became quite lucrative; and a version of *It's A Wonderful Life* can always be found on a local TV station around the holidays.

Home is Where the Future Is

While theaters will remain an important part of the movie business, home viewing is becoming the prevalent venue for Americans. Specifically, in a national survey conducted by the Consumer Electronics Manufacturers Association (CEMA), 64 percent of home theater owners said watching a movie at home "is more fun" than going out to the theater. In addition, according to CEMA's research, a majority of home theater owners say their systems' picture and sound quality is better than movie theaters!

In 1994, 473 million prerecorded cassettes were sold to U.S. dealers, according to the Motion Picture Association of America. The worldwide home video rental market reached 94 million units in 1994, according to Cambridge Associates.

Although the movie industry is making many advances in sound, animation, and other production techniques, the most significant development will come in the delivery of movies to the home. Satellite systems now offer digital quality movies to homes across the country. Telecommunication and cable companies are introducing video-on-demand systems that will allow consumers to order specific movies to be transmitted into the home at the time the viewer selects, and new video disc technologies offer the promise of low-cost digital movies for home use.

Appendix 1
FEWER CHIPS, LOWER COST, HIGHER RELIABILITY

Component manufacturers are working to support the new generation of HDTVs. An example of what is going on in the industry is the announcement by Philips Semiconductors of what it terms to be the industry's first two-chip hybrid analog and digital TV (DTV) front-end solution (*Figure A1.1*), the TDA8980 input processor and TDA8961 DTV demodulator/decoder, that supports both ATSC Vestigial SideBand (VSB) and NTSC channel decoding. Designed for use in high-definition TV (HDTV), set-top box (STB), and multimedia PC applications, the chipset offers designers of DTV and receiver modules a complete front-end receiver solution for a range of broadcast signals for North American and other ATSC-based markets.

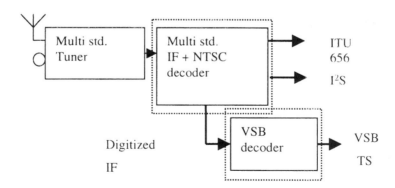

Philips Semiconductors

Figure A1.1
This chipset delivers a highly integrated, high-performance DTV front-end chipset with only two ICs. The solution offers both analog and digital terrestrial channel decoding capability, with comprehensive analog audio and video source selection capabilities. The chipset also supports format conversion for a variety of commonly used analog devices such as VCRs, cable TV and set-top boxes.

Earlier chipsets for this purpose required seven to nine integrated circuits (*Figure A1.2*), and so this second-generation chipset significantly reduces the chip count necessary to support these applications, while delivering outstanding reception at an attractive price. Furthermore, this solution reduces the number of interconnections required when using multiple front-end ICs, and therefore enhances the reliability of DTV reception performance.

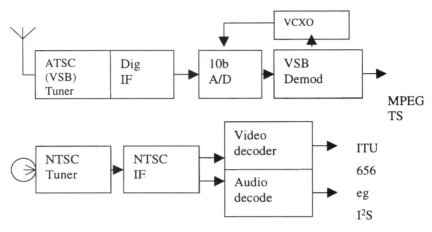

Philips Semiconductors

Figure A1.2
First-generation HDTV front-end solutions required seven to nine chips.

INCREASED INTEGRATION AND PERFORMANCE

Designed specifically to reduce the chip count from first-generation, front-end receiver solutions, this chipset delivers a highly integrated, high-performance DTV front-end chipset with only two ICs. The solution offers both analog and digital terrestrial channel decoding capability, with comprehensive analog audio and video source selection capabilities. The chipset also supports format conversion for a variety of commonly used analog devices such as VCRs, cable TV, and set-top boxes. This range of compatibility is necessary to support the many potential input devices desired by consumers. Furthermore, the ability to process both analog and digital signals provides users with an easy transition to digital content while maintaining access to traditional media sources.

ABOUT THE TWO-CHIP SOLUTION

The highly integrated solution features the TDA8980 input processor and a specialized TDA8961 VSB decoder (*Figure A1.3*). The advanced input processor eliminates many components by integrating NTSC decoding, an IF circuit, picture and sound demodulation, and VSB down conversion into a single device.

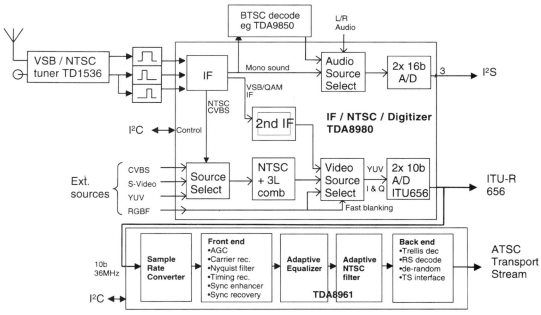

Figure A1.3
This two-chip HDTV front end packs a lot of circuitry into just two ICs.

The TDA8961 8-VSB demodulator/decoder IC offers intelligent features to retrieve signals despite man-made or naturally occurring echoes for outstanding reception performance. Additionally, the overall DTV performance is significantly enhanced through a unique cochannel filter that reduces interference from strong NTSC signals. The chip also supports parallel (8-bit) or serial MPEG2 transport stream output and includes an I^2C-bus interface.

Design Challenges of ATSC (VSB) Terrestrial Reception

Design Challenge	Cause	Where Found	Viewer Experience
Dynamic and Strong Static Multipath	Multiple reflections from high-rise buildings, autos, planes, etc.	Dense urban environments, especially using indoor antennas. Very common	Screen goes black
Poor/Variable Signal Strength	Natural, man-made obstacles between transmitter and receiver	Rural environments, indoor reception everywhere	Screen goes black intermittently
NTSC Co-channel and Adjacent Channel Interference	FCC frequency allocations, lower power (12 dB) of digital transmitters	Regional. Depends on frequency table and reception on fringes. Quite common.	Screen goes black
Long Static Echoes	Natural, man-made geographical features (usually isolated).	Parts of major cities, e.g. Chicago, Washington, D.C. Low percentage of overall sites.	Screen goes black

Table A1.1
A lot of things can happen to degrade a terrestrial signal. These are the challenges that designers had to overcome in designing the front end of an HDTV.

Appendix 2
HDTV EXAMPLE: HITACHI SDX SERIES

The information, specifications, glossary, and illustrations presented here are intended to give an idea of the kinds of features offered by digital television sets. Hitachi has introduced two DTV Models which are HDTV capable with a Set-Top Box. In other words, they are only missing the HDTV front end, which does the tuning and demodulation. These sets are the Series SDX. One is a 36-inch Direct View model and the other is a 60-inch projection model. A look at the specifications and the diagrams presented here gives an idea what HDTV is all about. Both of these units are feature-rich and are designed to be "interoperable" with computer systems. The information in this appendix was provided by Hitachi. Some of it may duplicate material found elsewhere in the book, but much of the information in this appendix provides additional insights into the world of HDTV.

SDX- HDTV READY

SDX series are available in both 60-inch projection and 36-inch direct view models. All time Progressive Scanning utilizing Hitachi's exclusive Flex Converter Technology converts all SDTV and NTSC signals into 480p, with no scan lines and no flicker. It is simply necessary to add an HDTV set-top box and these sets can provide True HDTV 1080i resolution. The 605DX88B utilizes the same optics as the HDTV model. The 365DX88B utilizes a high resolution Digital Pitch Invar picture tube and has an SVGA input for True SVGA Computer Resolution with PIPC (Picture in Picture-PC).

Digital Television Signals will be available to all of us through:

1. Over-the-air signal either with rooftop antennas or, if the signal is strong enough, through indoor antennas.

2. Cable companies, which are required to provide viewers with the entire 6 Mhz digital channel used by broadcasters, delivering all the broadcaster's digital bits intact.

3. Satellite Dish Receiver: Direct broadcast satellite systems offer consumers a second option for receiving high-definition digital TV signals. SAT provider Direct TV has announced plans for at least two high-definition channels, and upstart hardware and programming company Unity Motion hopes to make its mark both in the hardware and the software markets. In order to receive the HDTV channels, SAT subscribers will need a new elliptical dual-LNB dish that is capable of pointing at existing satellites as well as the Galaxy III-R satellite owned by Hughes sister company Panamstat, which Direct TV has an option to lease for expanded HDTV programming down the road.

In addition to excellent picture, DTV will also provide the best audio reproduction possible. With DTV technology consumers will enjoy the rich and true experience of theater-like sound with its 5.1 Dolby Digital broadcast signal. This signal will deliver a surround experience equivalent to what most DVD video disks deliver today, according to Dolby Labs.

FEATURES AND SPECIFICATIONS FOR HITACHI DTV

36SDX88B

Picture: 800 Lines of Resolution - SVGA 800x600 lines Computer Resolution - 1080i HDTV Capable with Set Top Box - Ultra Black (.78 mm) Digital Pitch Invar Picture Tube - Digital 3D Y/C Comb Filter - Advance Velocity Scan Modulation - Dynamic Focus

Sound: 10 Watts Front Firing Speaker System - MTS Stereo I SAP w/dbx - SRS Surround Sound - BBE Sound Clarity Enhancer - 7 Band Graphic Equalizer - Dynamic Bass - Loudness/Volume Correction - Loudness - Auto Noise Canceler.

Convenience: Advanced PIPC (Picture in Picture PC) - Genius MMV Illuminated Universal Remote Control - Ultra Tec Bit-Map On-Screen Display - 181 Channel Tuner with Auto/Programming - Channel ID -Video ID - Child Lock - Advanced Auto Demo - On-Screen Help - Improved Channel skip - Special Event Reminder - Sleep Timer - Favorite Channel (Family Favorites) - Power Resume - Channel Timer - MultiColor Plug and Play Interactive Set-Up - Advanced Auto Link - 4 Event On Off.

Specifications:

Screen Size: 36 V (measured diagonally)

Inputs:

Power Input	120 Volts AC	Component Video Input	
Power Consumption	180 Watts, Max 240	Luminance (Y)	1.0 Volt P-P, 75 Ohm
Antenna input impedance	75 Ohms	Chrominance (Cb, Cr)	0.7 Volt P-P, 75 Ohm
Channel Coverage	181 Channels	Audio Input Level (avg.)	400 mVrms, 47 K Ohm
VHF Band	2-13		
UHF Band	14-69		

Outputs:

Audio Output (variable)	400mVrms, I K Ohm
Video Output	1.0 Volt P-P, 75 Ohm

CATV Mid	A-5 — A-l
CATV Super Band	J-W
CATV Hiper Band	W+1 — W+28
CATV Ultra Band	W+29 — W+84
Video Input	1.0 Volt P-P, 75 Ohm

Dimensions:

Height	29 17/32 inches
Width	35 inches
Depth	24 13/32 inches
Weight	164 lbs.

S-Video Input
  Luminance (Y)   1.0 Volt P-P, 75 Ohm
  Chrominance (C)   0.286 Volt P-P, 75 Ohm

VGA and SVGA Specifications

Type	VGA1 I PS2-i	VGA2 I P52-2	SVGA *I* P52-3
Mode	640x480	640x480	800x600
Horizontal Frequency	31.469Khz	37.861 Khz	37.879 Khz
Vertical frequency	60.00 Hz	72.809 Hz	60.31 Hz
Sync type	H/V separate	H/V separate	H/V separate
Sync Polarity (H)	TTL: negative	TTL: positive	TTL: positive
(V)	TTL: negative	TTL: positive	TTL: positive

36SDX88B

FRONT PANEL CONTROLS

FRONT VIEW

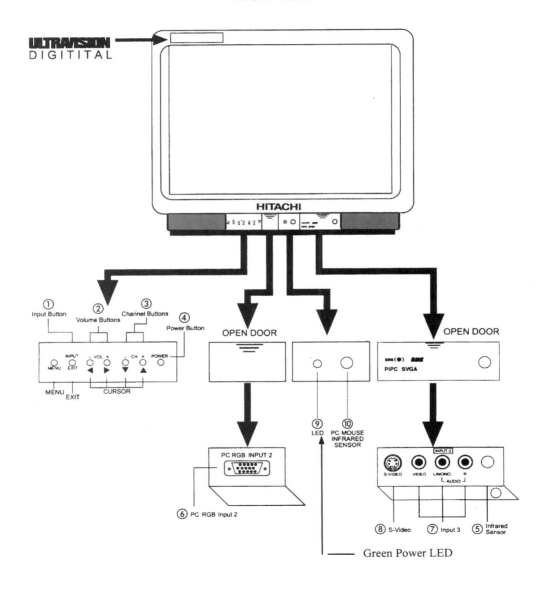

36SDX88B

ULTRATEC ON-SCREEN DISPLAY

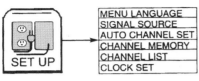

MENU LANGUAGE	Choose English, French, or Spanish language.
SIGNAL SOURCE	Select Antenna (Indoor or Outdoor) or Cable TV.
AUTO CHANNEL SET	First time set up for channel buttons.
CHANNEL MEMORY	Add or erase channel manually.
CHANNEL LIST	Check channel name, scan, and child lock.
CLOCK SET	Set before using timer features.

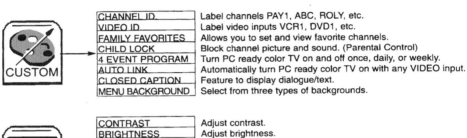

CHANNEL ID	Label channels PAY1, ABC, ROLY, etc.
VIDEO ID	Label video inputs VCR1, DVD1, etc.
FAMILY FAVORITES	Allows you to set and view favorite channels.
CHILD LOCK	Block channel picture and sound. (Parental Control)
4 EVENT PROGRAM	Turn PC ready color TV on and off once, daily, or weekly.
AUTO LINK	Automatically turn PC ready color TV on with any VIDEO input.
CLOSED CAPTION	Feature to display dialogue/text.
MENU BACKGROUND	Select from three types of backgrounds.

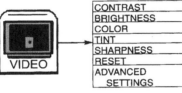

CONTRAST	Adjust contrast.
BRIGHTNESS	Adjust brightness.
COLOR	Adjust color.
TINT	Adjust tint.
SHARPNESS	Adjust sharpness.
RESET	Set VIDEO settings to factory preset condition.
ADVANCED SETTINGS	Improve picture performance and set component.

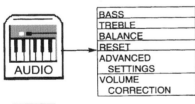

BASS	Adjust bass.
TREBLE	Adjust treble.
BALANCE	Adjust balance.
RESET	Set AUDIO settings to factory preset.
ADVANCED SETTINGS	Improve sound performance.
VOLUME CORRECTION	Lower volume on selected channels.

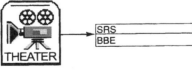

SRS	Produces 3-D sound using the PC ready color TV's speakers.
BBE	Produces High Definition sound that is natural and crisp using the PC ready color TV's speakers.

SPECIAL EVENT REMINDER	Set PC ready color TV to remind you of birthdays, etc..
CALENDAR	Check day, month, year, and special events.
AUTO HELP	Displays a description of the displayed menu.

REAR PANEL CONNECTIONS

36SDX88B

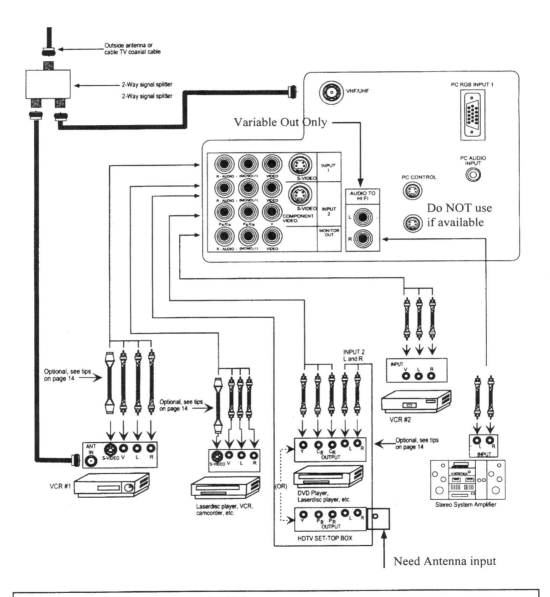

Outside antenna or cable TV coaxial cable

2-Way signal splitter
2-Way signal splitter

Variable Out Only

VHF/UHF

PC RGB INPUT 1

PC AUDIO INPUT

PC CONTROL

Do NOT use if available

AUDIO TO HI FI

L

R

R - AUDIO - (MONO) / L   VIDEO   S-VIDEO   INPUT 1

R - AUDIO - (MONO) / L   VIDEO   S-VIDEO   INPUT 2

COMPONENT VIDEO

P$_B$/C$_B$ P$_R$/C$_R$ Y

MONITOR OUT

R - AUDIO - (MONO) / L   VIDEO

Optional, see tips on page 14

Optional, see tips on page 14

INPUT 2 L and R

INPUT
V   L   R

VCR #2

Optional, see tips on page 14

INPUT
L   R

ANT IN
S-VIDEO V   L   R

VCR #1

S-VIDEO V   L   R

Laserdisc player, VCR, camcorder, etc.

Y   C$_B$ C$_R$ L   R
OUTPUT

(OR)

DVD Player, Laserdisc player, etc.

Y   B   P$_R$ L   R
OUTPUT

HDTV SET-TOP BOX

Stereo System Amplifier

Need Antenna input

NOTE: Connect only 1 component to each input jack.

36SDX88B

USING THE REMOTE TO CONTROL YOUR VCR FUNCTIONS

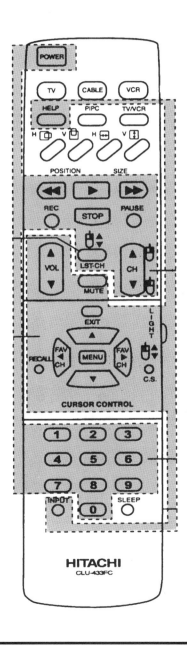

36SDX88B

PICTURE-IN-PC

Your PC ready color TV incorporates PinPC technology designed for improved viewing enjoyment.

The Picture-in-PC feature is convenient when you want to watch a program. You can watch your PC operation while viewing other programs from Antenna, Video and component input sources.

PC input sources can only be viewed as a main pictures.

Antenna, Video and component inputs can be viewed as the sub-picture.

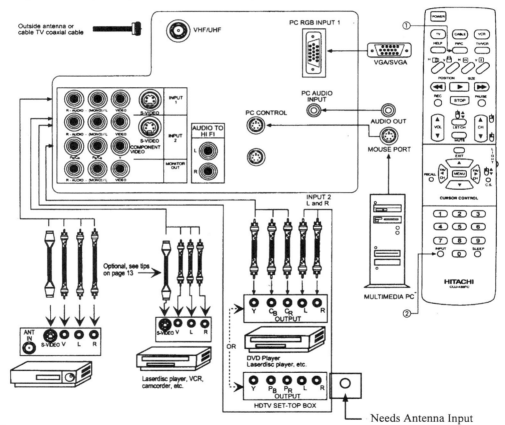

① **PiPC Button**
Press the PiPC Button and a Sub-Picture will appear. This PiPC works only on PC mode.

② **Input Button**
Press Input to change Sub-Picture (See Page 24) Input Source.

NOTE: When PiPC is on Antenna Source, press channel ▲, ▼ to change channel or use number button.

60SDX88B

Picture: 480 P Progressive Scan (Exclusive Flex Scan Coverter Double Scan Technology) - 1080i HDTV Capable with Set Top Box - Exclusive Ultra Focus SB Lens System and PRT's - Exclusive Magic Focus - High Spec Chassis - Digital 3D Y/C Comb Filter - Built-in High Contrast Ultra Shield - Digital A.I. Home Theater Modes - Color Purity Lens Filters (Red and Green) - First Surface Mirrors - Ultra Black Screen with Surface Diffuser - Advance Velocity Scan Modulation - Dynamic Focus.

Sound: MTS Stereo/SAP w/dbx - Dolby Pro Logic Surround Sound - 60 Watt Bass Reflex Speaker System - 2 Built-in Center Channel Speakers - 7 Band Graphic Equalizer - Dynamic Bass / Sub Woofer Control - Loudness I Perfect Volume - Wireless Sound Delivery System - Auto Noise Canceler.

Convenience: Multi Viewer 4 Dual Tuner Picture in Picture - Multi Page Genius Illuminated Universal Remote Control - Ultra Tec Bit-Map On-Screen Display - 181 Channel Tuner with Auto/Programming - Front Panel Menu Controls - Channel ID - Video ID - 4 Event On-Off Channel Timer - Child Lock - Front Panel Lock Out - Advanced Auto Demo - On Screen Help - Special Event Reminder - Sleep Timer - Favorite Channel (Family Favorites) - Power Resume -Channel Skip - Multi Color Plug and Play Interactive Set-Up -Advanced Auto Link - 2 Level Mute.

Specifications:

Screen Size: 60 inches (measured diagonally)

Inputs:

Power Input .... 120 Volts AC, 60 Hz

Power Consumption .... 245 Watts, Max 290

Antenna input impedance .... 75 Ohms

Channel Coverage .... 181 Channels

VHF Band 2-13

UHF Band 14-69

CATV Mid Band .... A-5 — A-1

CATV Super Band .... J-W

CATV Hiper Band .... W+1 — W+28

CATV Ultra Band .... W+29 — W+84

Video Input .... 1.0 Volt P-P, 75 Ohm

S-Video Input
    Luminance (Y) .... 1.0 Volt P-P, 75 Ohm
    Chrominance (C) .... 0.286 Volt P-P, 75 Ohm

Component Video Input
    Luminance (Y) .... 1.0 Volt P-P, 75 Ohm
    Chrominance (Cb, Cr) .... 0.7 Volt P-P, 75 Ohm

Audio Input Level (avg.) .... 400 mVrms, 47 K Ohm

Outputs:

Audio Output (variable) .... 400mVrms, 1 K Ohm

Video Output .... 1.0 Volt P-P, 75 Ohm

Dimensions:

Height .... 60 7/16 inches

Width .... 51 3/8 inches

Depth .... 27 1/2 inches

Weight .... 255 lbs.

60SDX88B

ULTRATEC OSD

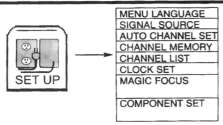

MENU LANGUAGE	Choose English, French, or Spanish text.
SIGNAL SOURCE	Select Antenna or Cable TV.
AUTO CHANNEL SET	First time set up for channel buttons.
CHANNEL MEMORY	Channel buttons, add, skip.
CHANNEL LIST	Check channel name, scan, and child lock.
CLOCK SET	Set before using timer features.
MAGIC FOCUS	Automatically match red, green, and blue colors to make white.
COMPONENT SET	Select Aspect Ratio and type of component input.

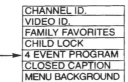

CHANNEL ID.	Label channels PAY1, ABC, etc.
VIDEO ID.	Label video inputs VCR1, DVD1, etc.
FAMILY FAVORITES	Allows you to set and view favorite channels.
CHILD LOCK	Block channel picture and sound.
4 EVENT PROGRAM	Turn TV on and off once, daily, or weekly.
CLOSED CAPTION	Feature to display dialogue/text.
MENU BACKGROUND	Select from three types of backgrounds.

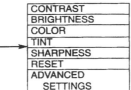

CONTRAST	Adjust contrast.
BRIGHTNESS	Adjust brightness.
COLOR	Adjust color.
TINT	Adjust tint.
SHARPNESS	Adjust sharpness.
RESET	Set VIDEO settings to factory preset.
ADVANCED SETTINGS	Improve picture performance.

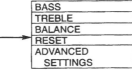

BASS	Adjust bass.
TREBLE	Adjust treble.
BALANCE	Adjust balance.
RESET	Set AUDIO settings to factory preset.
ADVANCED SETTINGS	Improve sound performance.

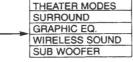

THEATER MODES	Picture and sound are automatically set.
SURROUND	Special sound effects, including Dolby Pro Logic.
GRAPHIC EQ.	More precise audio control.
WIRELESS SOUND	Select from three types of audio for wireless output.
SUB WOOFER	Adjust Sub-Woofer volume.

| SPECIAL EVENT REMINDER | Set TV to remind you of birthdays, etc.. |
| CALENDAR | Check day, month, year, and special events. |

60SDX88B

REAR PANEL JACKS

TYPICAL FULL-FEATURE SETUP

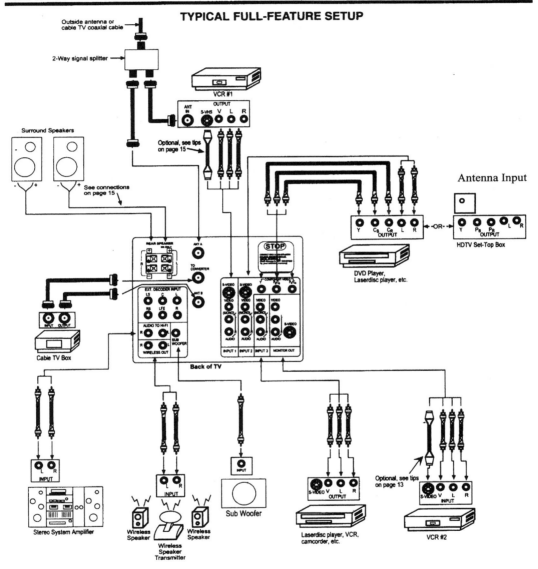

| NOTE: | Connect only 1 component to each input jack. |

60SDX88B Only
Remote Control Change

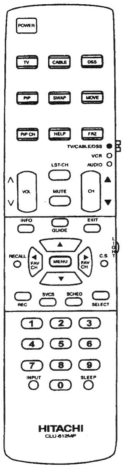

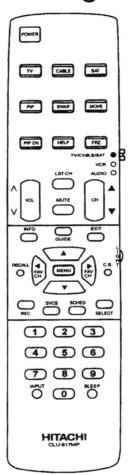

Remote Control provided
(CLU-612MP)

Remote Control as seen in the
Instruction Book
(CLU-617MP)

Shown above is the Remote Control that was provided with your projection television set. As you use this manual, whenever the "SAT" button is referenced, you will instead press the "DSS" button.

Shown above is the Remote Control that you will see referenced throughout this manual.

Appendix 3
DTV BROADCASTERS
START-UP SCHEDULE
(AS OF 3/1/1999)

Station	City	Date On	Owner	Format	Time of Day
WCBS (CBS)	New York	November 1	CBS	1080i	24 Hr
KCOP (UPN)	Los Angeles	December 15	Chris Craft	480p	24 Hr
KTLA (WB)	Los Angeles	October 29	Tribune	1080i	12 Hr
KCBS (CBS)	Los Angeles	November 1	CBS	1080i	24 Hr
KNBC (NBC)	Los Angeles	October 22	NBC	1080i	24 Hr
KABC (ABC)	Los Angeles	November 1	NA	NA	NA
WPVI (ABC)	Philadelphia	November 1	ABC	720p	24 Hr
KYW (CBS)	Philadelphia	November 1	CBS	1080i	24 Hr
WTXF (Fox)	Philadelphia	October 27	Fox	720p	24 Hr
WCAU (NBC)	Philadelphia	November 6	NBC	1080i	24 Hr
KGO (ABC)	San Francisco	November 18	ABC	NA	NA
KRON (NBC)	San Francisco	November 30	Chronicle	NA	NA
KPIX (CBS)	San Francisco	November 18	CBS	NA	NA
KTVU (Fox)	San Francisco	November 18	Cox	720p	24 Hr
WCVB (ABC)	Boston	October 30	Hearst-Argyle	720p	24 Hr
WMUR (ABC)	Boston	November 1	Imes	480i	
WHSH (HSN)	Boston	November 1	USA Broadcast	480p	8 Hr
WFAA (ABC)	Dallas	November 1	Belo	1080i	24 Hr
KDFW (Fox)	Dallas	November 1	Fox	720p	24 Hr
KXAS (NBC)	Dallas	November 1	NBC	1080i	24 Hr
WJLA (ABC)	Washington	November 1	Allbritton	480p	24 Hr

WUSA (CBS)	Washington	November 1	Gannett	NA	NA
WRC (NBC)	Washington	November 1	NBC	1080i	24 Hr
WETA (PBS)	Washington	November 9	Public	1080i	24 Hr
WXYZ (ABC)	Detroit	October 28	Scripps Howard	720p	17 Hr
WJBK (Fox)	Detroit	November 1	Fox	720p	24 Hr
WSB (ABC)	Atlanta	November 1	Cox	720p	24 Hr
WXIA (NBC)	Atlanta	November 1	Gannett	1080i	24 Hr
WGNX (CBS)	Atlanta	n/a	Trib/Meredith	1080i	24 Hr
KHOU (CBS)	Houston	November 1	Belo	1080i	24 Hr
KOMO (ABC)	Seattle	October 21	Fisher	720p	24 Hr
KCTS (PBS)	Seattle	September 18	Public	1080i	24 Hr
KING NBC)	Seattle	October 28	Belo	1080i	24 Hr
KATU (ABC)	Portland	October 26	Fisher	720p	24 Hr
KOPB (PBS)	Portland	November 9	Public	1080i	24 Hr
WTHR (NBC)	Indianapolis	September 28	Dispatch	1080i	24 Hr
WISH (CBS)	Indianapolis	December 16	LIN	1080i/480i	24 Hr
WBTV (CBS)	Charlotte	November 1	Jefferson-Pilot	1080i	24 Hr
WLWT (NBC)	Cincinnati	November 1	Gannett	1080i	24 Hr
WCPO (ABC)	Cincinnati	November 25	Scripps Howard	720p	10-6 M-F
WKRC (CBS)	Cincinnati	November 8	Jacor	1080i	Varies
WRAL (CBS)	Raleigh	May 7	Capitol	1080i	24 Hr
WBNS (CBS)	Columbus	November 1	Dispatch	NA	NA
WKOW (ABC)	Madison	November 1	Shockley	720p	18 Hr
WITF (ABC)	Honolulu	January	Hearst-Argyle	720p	24 Hr
WITF (PBS)	Harrisburg	November 9	Public	1080i	24 Hr
KCPT (PBS)	Kansas City	November 9	Public	1080i	24 Hr
WMPN (PBS)	Jackson	November 9	Public	1080i	24 Hr
WMVT (PBS)	Milwaukee	October	Public	Varies	NA
WTHN (ABC)	New Haven	December 4	LIN	720p	24 Hr

Note: NA = Not Available at the time this list was compiled

Appendix 4
DEFINITION OF TERMS AND SYMBOLS, ABBREVIATIONS, AND MATHEMATICAL OPERATORS

HDTV is a new adventure for programmers, broadcasters, and viewers alike. It is a complete departure from the familiar NTSC system, and understanding the concepts of HDTV requires a whole new vocabulary and a familiarity with the symbols, abbreviations, and mathematical operators that are used in the advanced TV system. This list will provide a quick reference for readers to turn to when they encounter a new and unfamiliar term.

DEFINITION OF TERMS

For the purposes of the Digital Television Standard, the following definition of terms apply:

8 VSB: Vestigial sideband modulation with 8 discrete amplitude levels.

8VSB Digital Decoder: Vestigial Sideband. New digital tuner will accept all 18 digital broadcast formats.

16 VSB: Vestigial sideband modulation with 16 discrete amplitude levels.

ACATS: Advisory Committee on Advanced Television Service.

Advanced Auto Link: Automatically turns on TV and switches to correct video input when an external component is activated.

Access Unit: A coded representation of a presentation unit. In the case of audio, an access unit is the coded representation of an audio frame. In the case of video, an access unit includes all the coded data for a picture, and any stuffing that follows it, up to but not including the start of the next access unit. If a picture is not preceded by a group_start_code or a sequence_header_code, the access unit begins with a the picture start code. If a picture is preceded by a group_start_code and/or a sequence_header_code, the access unit begins with the first byte of the first of these start codes. If it is the last picture preceding a sequence_end_code in the bit stream all bytes between the last byte of the coded picture and the sequence_end_code (including the sequence_end_code) belong to the access unit.

A/D: Analog to digital converter.

AES: Audio Engineering Society.

Anchor Frame: A video frame that is used for prediction. I-frames and P-frames are generally used as anchor frames, but B-frames are never anchor frames.

ANSI: American National Standards Institute.

Asynchronous Transfer Mode (ATM): A digital signal protocol for efficient transport of both constant-rate and bursty information in broadband digital networks. The ATM digital stream consists of fixed-length packets called "cells," each containing 53 8-bit bytes—a 5-byte header and a 48-byte information payload.

ATEL: Advanced Television Evaluation Laboratory.

ATM: See asynchronous transfer mode.

ATSC: Advanced Television Standards Committee. Group that defines the new digital TV standards.

ATTC: Advanced Television Test Center.

ATV: The U. S. advanced television system.

BBE Sound Clarity Enhancer: Re-creates the clear midrange and high sounds of the original recordings.

Bidirectional Pictures or **B-pictures** or **B-frames:** Pictures that use both future and past pictures as a reference. This technique is termed *bidirectional prediction*. B-pictures provide the most compression. B-pictures do not propagate coding errors as they are never used as a reference.

bit rate: The rate at which the compressed bit stream is delivered from the channel to the input of a decoder.

block: A block is an 8-by-8 array of pel values or DCT coefficients representing luminance or chrominance information.

bps: Bits per second.

byte-aligned: A bit in a coded bit stream is byte-aligned if its position is a multiple of 8 bits from the first bit in the stream.

CDTV: See conventional-definition television.

channel: A digital medium that stores or transports a digital television stream.

Child Lock System: Block out any unwanted channels. Child lock functions are accessed with a four-digit code of your choosing. (Use 7777 to clear)

coded representation: A data element as represented in its encoded form.

compression: Reduction in the number of bits used to represent an item of data.

constant bit rate: Operation where the bit rate is constant from start to finish of the compressed bit stream.

conventional-definition television (CDTV): This term is used to signify the *analog* NTSC television system as defined in ITU-R Recommendation 470. See also *standard-definition television* and ITU-R Recommendation 1125.

CRC: The cyclic redundancy check to verify the correctness of the data.

D-frame: Frame coded according to an MPEG-1 mode which uses DC coefficients only.

data element: An item of data as represented before encoding and after decoding.

DCT: See discrete cosine transform.

DDC2B Protocol: Automatically adjusts computer to proper resolution level when connected to a system running Windows 95 and above. Models 36SDX88B & 27MMV30B.

decoded stream: The decoded reconstruction of a compressed bit stream.

decoder: An embodiment of a decoding process.

decoding (process): The process defined in the Digital Television Standard that reads an input coded bit stream and outputs decoded pictures or audio samples.

decoding time-stamp (DTS): A field that may be present in a PES packet header that indicates the time that an access unit is decoded in the system target decoder.

digital storage media (DSM): A digital storage or transmission device or system.

Digital TV (DTV): This is the umbrella term used to describe the new digital television system adopted by the FCC in December 1996.

DIRECTV: System receiver for standard- and high-definition reception: Receive all digital and high-definition satellite broadcast programming anywhere in the country with the built-in High-Definition receiver.

discrete cosine transform: A mathematical transform that can be perfectly undone and which is useful in image compression.

Dolby Digital Surround Sound: Audio is re-created using a pure digital sound. Unlike Dolby Pro Logic which uses four channels of sound, Dolby uses 5.1 channels.

DSM-CC: Digital storage media command and control.

DSM: Digital storage media.

DTS: See decoding time-stamp.

DVCR: Digital video cassette recorder

ECM: See entitlement control message.

editing: A process by which one or more compressed bit streams are manipulated to produce a new compressed bit stream. Conforming edited bit streams are understood to meet the requirements defined in the Digital Television Standard.

elementary stream (ES): A generic term for one of the coded video, coded audio, or other coded bit streams. One elementary stream is carried in a sequence of PES packets with one and only one stream_id.

elementary stream clock reference (ESCR): A time stamp in the PES Stream from which decoders of PES streams may derive timing.

EMM: See entitlement management message.

encoder: An embodiment of an encoding process.

encoding (process): A process that reads a stream of input pictures or audio samples and produces a valid coded bit stream as defined in the Digital Television Standard.

entitlement control message (ECM): Entitlement control messages are private conditional access information which specify control words and possibly other stream-specific, scrambling, and/or control parameters.

entitlement management message (EMM): Entitlement management messages are private conditional access information which specify the authorization level or the services of specific decoders. They may be addressed to single decoders or groups of decoders.

entropy coding: Variable length lossless coding of the digital representation of a signal to reduce redundancy.

entry point: Refers to a point in a coded bit stream after which a decoder can become properly initialized and commence syntactically correct decoding. The first transmitted picture after an entry point is either an I-picture or a P-picture. If the first transmitted picture is not an I-picture, the decoder may produce one or more pictures during acquisition.

ES: See elementary stream.

ESCR: See elementary stream clock reference.

event: An event is defined as a collection of elementary streams with a common time base, an associated start time, and an associated end time.

field: For an interlaced video signal, a "field" is the assembly of alternate lines of a frame. Therefore, an interlaced frame is composed of two fields, a top field and a bottom field.

forbidden: This term, when used in clauses defining the coded bit stream, indicates that the value shall never be used. This is usually to avoid emulation of start codes.

FPLL: Frequency and phase-locked loop.

First Surface Mirror: Light reflects off the front surface of the mirror eliminating the minor distortion caused by light passing through the glass to the reflective surface and then back out.

frame: A frame contains lines of spatial information of a video signal. For progressive video, these lines contain samples starting from one time instant and continuing through successive lines to the bottom of the frame. For interlaced video, a frame consists of two fields, a top field and a bottom field. One of these fields will commence one field later than the other.

GOP: See group of pictures.

Group of pictures (GOP): A group of pictures consists of one or more pictures in sequence.

High-Definition Component Video Inputs: Achieves the highest possible resolution from a DVD or digital set-top box. The video signal is split into luminance (BIW) and chrominance (color), which is then further separated into blue (Pb) and red (Pr) components. Unlike conventional component inputs (Y, Cb and Cr) that can only accept 480i signals, SDX high resolution inputs (Y, Pb and Pr) will accept 480P and full 1080i HDTV signals.

High-Definition Television (HDTV): This term refers to a complete product/system with the following minimum performance attributes:

Resolution: Vertical resolution of 720P, 1080i or higher.

Aspect Ratio: Capable of dispaying a 16x9 image at the minimum resolution level.

Audio: Receives, reproduces, and/or outputs Dolby Digital Audio.

Receiver: Receives all ATSC Table III Formats.

high-definition television (HDTV): High-definition television has a resolution of approximately twice that of conventional television in both the horizontal (H) and vertical (V) dimensions and a picture aspect ratio (HxV) of 16:9. ITU-R Recommendation 1125 further defines "HDTV quality" as the delivery of a television picture which is subjectively identical with the interlaced HDTV studio standard.

high level: A range of allowed picture parameters defined by the MPEG-2 video coding specification which corresponds to high definition television.

High Spec Chassis: Ten advanced picture quality circuits and two high-resolution inputs. Features include:

Automatic Brightness Limiter: Controls maximum brightness levels to avoid blooming.

Digital Comb Filter, 3D Y/C type: Uses current and previous video frame information to further eliminate picture artifacts.

Digital Convergence, Magic Focus: Automatic digital convergence. Aligns *256* convergence points (x3 PRT's) across the screen. Magic Focus is the most accurate convergence system available.

Dynamic Black Enhancer: Amplifies the luminance signal in the dark areas of the picture to increase picture detail.

Dynamic Noise Reducer: Reduces noise without decreasing the sharpness of the brighter areas.

Fresh Green Circuit: Creates true vivid greens.

Signal Tracker: Controls color saturation and reproduces true flesh tones.

T. C. P. E.: Improves dark to light scene transitions. Improves picture sharpness.

Vertical Enhancer: Uses information from the Digital Comb Filter to enhance horizontal edges.

Wide Band Video Amplifier: Passes higher frequencies to increase resolution and clarity.

S-Video Input: Used for high-resolution sources such as Super VHS Video Cassette Players, Camcorders, or any other video source with S-Video output capability.

Huffman coding: A type of source coding that uses codes of different lengths to represent symbols which have unequal likelihood of occurrence.

IEC: International Electrotechnical Commission.

Interlaced Scan: Two fields of video information are interlaced to form one video frame.

intra coded pictures or **I-pictures** or **I-frames:** Pictures that are coded using information present only in the picture itself and not depending on information from other pictures. I-pictures provide a mechanism for random access into the compressed video data. I-pictures employ transform coding of the pel blocks and provide only moderate compression.

ISO: International Organization for Standardization.

ITU: International Telecommunication Union.

JEC: Joint Engineering Committee of EIA and NCTA.

layer: One of the levels in the data hierarchy of the video and system specification.

level: A range of allowed picture parameters and combinations of picture parameters.

macroblock: In the ATV system, a macroblock consists of four blocks of luminance and one each Cr and Cb block.

main level: A range of allowed picture parameters defined by the MPEG-2 video coding specification with maximum resolution equivalent to ITU-R Recommendation 601.

main profile: A subset of the syntax of the MPEG-2 video coding specification that is expected to be supported over a large range of applications.

Mbps: 1 million bits per second.

motion vector: A pair of numbers that represent the vertical and horizontal displacement of a region of a reference picture for prediction.

MP@HL: Main profile at high level.

MP@ML: Main profile at main level.

MPEG: Refers to standards developed by the ISO/IEC JTC1/SC29 WG11, *Moving Picture Experts Group.* MPEG may also refer to the Group.

MPEG-1: Refers to ISO/IEC standards 11172-1 (Systems), 11172-2 (Video), 11172-3 (Audio), 11172-4 (Compliance Testing), and 11172-5 (Technical Report).

MPEG-2: Refers to ISO/IEC standards 13818-1 (Systems), 13818-2 (Video), 13818-3 (Audio), 13818-4 (Compliance).

NTSC: National Television Standards Committee. Current analog broadcast television standards.

pack: A pack consists of a pack header followed by zero or more packets. It is a layer in the system coding syntax.

packet data: Contiguous bytes of data from an elementary data stream present in the packet.

packet identifier (PID): A unique integer value used to associate elementary streams of a program in a single or multi-program transport stream.

packet: A packet consists of a header followed by a number of contiguous bytes from an elementary data stream. It is a layer in the system coding syntax.

padding: A method to adjust the average length of an audio frame in time to the duration of the corresponding PCM samples, by continuously adding a slot to the audio frame.

payload: Payload refers to the bytes which follow the header byte in a packet. For example, the payload of a transport stream packet includes the PES_packet_header and its PES_packet_data_bytes or pointer_field and PSI sections, or private data. A PES_packet_payload, however, consists only of PES_packet_data_bytes. The transport stream packet header and adaptation fields are not payload.

PCR: See program clock reference.

pel: See pixel.

PES packet header: The leading fields in a PES packet up to, but not including, the PES_packet_data_byte fields where the stream is not a padding stream. In the case of a padding stream, the PES packet header is defined as the leading fields in a PES packet up to but not including the padding_byte fields.

PES packet: The data structure used to carry elementary stream data. It consists of a packet header followed by PES packet payload.

PES Stream: A PES stream consists of PES packets, all of whose payloads consist of data from a single elementary stream, and all of which have the same stream_id.

PES: An abbreviation for packetized elementary stream.

picture: Source, coded, or reconstructed image data. A source or reconstructed picture consists of three rectangular matrices representing the luminance and two chrominance signals.

PID: See packet identifier.

pixel: "Picture element" or "pel." A pixel is a digital sample of the color intensity values of a picture at a single point.

predicted pictures or **P-pictures** or **P-frames:** Pictures that are coded with respect to the nearest *previous* I- or P-picture. This technique is termed *forward prediction*. P-pictures provide more compression than I-pictures and serve as a reference for future P-pictures or B-pictures. P-pictures can propagate coding errors when P-pictures (or B-pictures) are predicted from prior P-pictures where the prediction is flawed.

presentation time-stamp (PTS): A field that may be present in a PES packet header that indicates the time that a presentation unit is presented in the system target decoder.

presentation unit (PU): A decoded audio access unit or a decoded picture.

profile: A defined subset of the syntax specified in the MPEG-2 video coding specification

program clock reference (PCR): A time stamp in the transport stream from which decoder timing is derived.

program element: A generic term for one of the elementary streams or other data streams that may be included in the program.

program specific information (PSI): PSI consists of normative data which is necessary for the demultiplexing of transport streams and the successful regeneration of programs.

PIPC: Picture-in-Picture-PC. View video from any source in the subpicture while using the TV as a computer monitor. Model 365DX88B

program: A program is a collection of program elements. Program elements may be elementary streams. Program elements need not have any defined time base; those that do have a common time base and are intended for synchronized presentation.

Progressive Scan: Non-Interlaced scanning process. Complete video frames are transmitted at one time. Eliminates scan lines and flicker, same scanning used on computer monitors.

PSI: See program specific information.

PTS: See presentation time-stamp.

PU: See presentation unit.

quantizer: A processing step which intentionally reduces the precision of DCT coefficients

random access: The process of beginning to read and decode the coded bit stream at an arbitrary point.

reserved: This term, when used in clauses defining the coded bit stream, indicates that the value may be used in the future for Digital Television Standard extensions. Unless otherwise specified within this Standard, all reserved bits shall be set to "1".

SCR: See system clock reference.

scrambling: The alteration of the characteristics of a video, audio, or coded data stream in order to prevent unauthorized reception of the information in a clear form. This alteration is a specified process under the control of a conditional access system.

SDTV: See standard-definition television.

slice: A series of consecutive macroblocks.

SMPTE: Society of Motion Picture and Television Engineers.

source stream: A single, nonmultiplexed stream of samples before compression coding.

splicing: The concatenation performed on the system level or two different elementary streams. It is understood that the resulting stream must conform totally to the Digital Television Standard.

Standard-Definition Television (SDTV): This term refers to a complete product/system with the following performance attributes:

Resolution: Display resolution lower than that of IIDTV.

Aspect Ratio: None specified.

Audio: Produces usable audio.

Receiver: Receives all ATSC Table III Formats, and produces a usable picture.

standard-definition television (SDTV): This term is used to signify a *digital* television system in which the quality is approximately equivalent to that of NTSC. This equivalent quality may be achieved from pictures sourced at the 4:2:2 level of ITU-R Recommendation 601 and subjected to processing as part of the bit rate compression. The results should be such that when judged across a representative sample of program material, subjective equivalence with NTSC is achieved. Also called standard digital television. See also *conventional-definition television* and ITU-R Recommendation 1125.

start codes: 32-bit codes embedded in the coded bit stream that are unique. They are used for several purposes including identifying some of the layers in the coding syntax. Start codes consist of a 24-bit prefix (0x000001) and an 8-bit stream_id.

STD input buffer: A first-in, first-out buffer at the input of a system target decoder for storage of compressed data from elementary streams before decoding.

STD: See system target decoder.

still picture: A coded still picture consists of a video sequence containing exactly one coded picture which is intracoded. This picture has an associated PTS and the presentation time of succeeding pictures, if any, is later than that of the still picture by at least two picture periods.

system clock reference (SCR): A time stamp in the program stream from which decoder timing is derived.

system header: The system header is a data structure that carries information summarizing the system characteristics of the Digital Television Standard multiplexed bit stream.

system target decoder (STD): A hypothetical reference model of a decoding process used to describe the semantics of the Digital Television Standard multiplexed bit stream.

time-stamp: A term that indicates the time of a specific action such as the arrival of a byte or the presentation of a presentation unit.

TOV: Threshold of visibility.

transport stream packet header: The leading fields in a transport stream packet up to and including the continuity_counter field.

UltraBlack High Contrast Screen: Enhances and sharpens the picture using a .72 mm fine pitch screen. By concentrating the light diffusion layer (tint) at the front surface of the screen, reflected ambient light is reduced while increasing contrast over 30 percent. In addition, color shift is reduced and the viewing angle is increased to 160 degrees.

UltraTEC Bit Map OSD: Multi-Color, bit map graphics with "Windows"-style drop-down menus make this the simplest operating menu system available.

variable bit rate: Operation where the bit rate varies with time during the decoding of a compressed bit stream.

VBV: See video buffering verifier.

Video buffering verifier (VBV): A hypothetical decoder that is conceptually connected to the output of an encoder. Its purpose is to provide a constraint on the variability of the data rate that an encoder can produce.

video sequence: A video sequence is represented by a sequence header, one or more groups of pictures, and an end_of_sequence code in the data stream.

SYMBOLS, ABBREVIATIONS, AND MATHEMATICAL OPERATORS

The symbols, abbreviations, and mathematical operators used to describe the Digital Television Standard are those adopted for use in describing MPEG-2 and are similar to those used in the "C" programming language. However, integer division with truncation and rounding are specifically defined. The bitwise operators are defined assuming two's-complement representation of integers. Numbering and counting loops generally begin from 0.

Arithmetic Operators

+   Addition.

-          Subtraction (as a binary operator) or negation (as a unary operator).

++   Increment.

- -   Decrement.

* or ´   Multiplication.

^   Power.

/   Integer division with truncation of the result toward 0. For example, 7/4 and -7/-4 are truncated to 1 and -7/4 and 7/-4 are truncated to -1.

//   Integer division with rounding to the nearest integer. Half-integer values are rounded away from 0 unless otherwise specified. For example 3//2 is rounded to 2, and -3//2 is rounded to -2.

DIV   Integer division with truncation of the result towards -¥.

%   Modulus operator. Defined only for positive numbers.

Sign( ) $Sign(x) = 1 \quad x > 0$

$$0 \quad x == 0$$

$$-1 \quad x < 0$$

NINT ( )   Nearest integer operator. Returns the nearest integer value to the real-valued argument. Half-integer values are rounded away from 0.

sin   Sine.

cos   Cosine.

exp   Exponential.

Ö   Square root.

$\log_{10}$ Logarithm to base ten.

$\log_e$ Logarithm to base e.

Logical Operators

|| Logical OR.

&&   Logical AND.

!   Logical NOT.

Relational Operators

>   Greater than.

3   Greater than or equal to.

<   Less than.

£   Less than or equal to.

==   Equal to.

!=   Not equal to.

max [,....,]   The maximum value in the argument list.

min [,....,]   The minimum value in the argument list.

Bitwise Operators

    &   AND.

    | OR.

    >>   Shift right with sign extension.

    <<   Shift left with 0 fill.

Assignment

    =   Assignment operator.

MNEMONICS

The following mnemonics are defined to describe the different data types used in the coded bit stream.

    bslbf   Bit string, left bit first, where "left" is the order in which bit strings are written in the Standard. Bit strings are written as a string of 1s and 0s within single quote marks, e.g. '1000 0001'. Blanks within a bit string are for ease of reading and have no significance.

    uimsbf Unsigned integer, most significant bit first.

The byte order of multi-byte words is most significant byte first.

CONSTANTS

    p   3.14159265359...

    e   2.71828182845...

Method of Describing Bit Stream Syntax

Each data item in the coded bit stream described below is in bold type. It is described by its name, its length in bits, and a mnemonic for its type and order of transmission.

The action caused by a decoded data element in a bit stream depends on the value of that data element and on data elements previously decoded. The decoding of the data elements and definition of the state variables used in their decoding are described in the clauses containing the semantic description of the syntax. The following constructs are used to express the conditions when data elements are present, and are in normal type.

Note this syntax uses the "C" code convention that a variable or expression evaluating to a non-zero value is equivalent to a condition that is true.

while (condition) {
 	
data_element
 	
. . .
 	
}
do{
	If the condition is true, then the group of data elements occurs next in the data stream. This repeats until the condition is not true.

data_element . . . } while (condition)	The data element always occurs at least once. The data element is repeated until the condition is not true.

if (condition) {
 	
data_element
 	
. . .
 	
}
	If the condition is true, then the first group of data elements occurs next in the data stream.

else {
 	
data_element
 	
. . .
 	
}
	If the condition is not true, then the second group of data elements occurs next in the data stream.

for (i = 0;i<n;i++) {

data_element

. . .

}

The group of data elements occurs n times. Conditional constructs within the group of data elements may depend on the value of the loop control variable i, which is set to zero for the first occurrence, incremented to 1 for the second occurrence, and so forth.

As noted, the group of data elements may contain nested conditional constructs. For compactness, the {} are omitted when only one data element follows.

data_element [ ]

data_element [ ] is an array of data. The number of data elements is indicated by the context.

data_element [n]

data_element [n] is the n+1th element of an array of data.

data_element [m][n]

data_element [m][n] is the m+1,n+1 th element of a two-dimensional array of data.

data_element

[l][m][n]

data_element [l][m][n] is the l+1,m+1,n+1 th element of a three-dimensional array of data.

data_element [m..n]

data_element [m..n] is the inclusive range of bits between bit m and bit n in the data_element.

Decoders must include a means to look for start codes and sync bytes (transport stream) in order to begin decoding correctly, and to identify errors, erasures or insertions while decoding. The methods to identify these situations, and the actions to be taken, are not standardized.

Definition of Byte-aligned Function

The function byte-aligned ( ) returns 1 if the current position is on a byte boundary; that is, the next bit in the bit stream is the first bit in a byte. Otherwise it returns 0.

Definition of Nextbits Function

The function nextbits( ) permits comparison of a bit string with the next bits to be decoded in the bit stream.

Definition of next_start_code Function

The next_start_code( ) function removes any zero bit and zero byte stuffing and locates the next start code.

This function checks whether the current position is byte-aligned. If it is not, 0 stuffing bits are present. After that any number of 0 bytes may be present before the start-code. Therefore start-codes are always byte-aligned and may be preceded by any number of 0 stuffing bits.

Table 3.1 Next Start Code

Syntax	No. of bits	Mnemonic c

next_start_code( ) {
    while (!byte-aligned( ))
        zero_bit
    while (nextbits( )!='0000 0000 0000 0000 0000 0001')
        zero_byte
}
	
	1
	
	8
		
		'0'
		
		'00000000'

Appendix 5
ANTENNAS FOR HDTV

Because of a number of factors, including the reluctance on the part of cable TV services to handle HDTV programming, at least in the short term, large numbers of viewers will be receiving their HDTV signals over the air via terrestrial broadcast from familiar local TV broadcasting stations. That means that consumers will have to have TV broadcast reception antennas.

The nature of HDTV is such that if there is a problem with reception of the signal, the consumer may not be able to display a picture at all. In the NTSC system, if a signal is not of sufficient strength, or if there is multipath reception, the displayed image will deteriorate, but the receiving set will still continue to show a picture. With HDTV, if one of these problems occurs, there will be no display. In the HDTV system, either the picture is perfect, or it just is not there.

It is, therefore, important that consumers have an antenna that is adequate to receive the HDTV signal based on their location with respect to the broadcast station, and on the unique broadcast reception signals that exist in that area. The Consumer Electronics Manufacturers Association has taken great pains to see that consumers have all the information necessary to select an adequate antenna.

THE NEED FOR AN INDUSTRY SOLUTION

The CEMA Antenna Mapping Program is the result of intense cross-industry efforts to provide antenna solutions for retail sales professionals. The first Antenna Summit, held in 1997, was held because research conducted by U.S. Satellite Broadcasting that revealed that 86 percent of small satellite nonbuyers didn't purchase a system because there was no satisfactory solution presented to them on how to receive local channels. (Source: Frank N. Magid and Associates, Inc., 6/96)

In October 1997, the CEMA Antenna Subdivision was formed. The group was formed to break through the sales barrier created by the cable industry that told consumers local channel reception was unavailable to them with DBS systems.

A second antenna summit was held in November 1997, and participants included TV antenna manufacturers, broadcasters, satellite program providers, and consumer electronics manufacturers. The goal was to explore concrete ways to provide retail sales people with greater sales success by creating a user-friendly method of recommending the right antenna for every customer.

In order to estimate antenna requirements, a variety of new technologies was utilized in the process. Factors taken into consideration in the creation of the antenna maps include:

• Height and location of local broadcast towers

• Broadcast frequencies

• Power of broadcast stations

• Topographical information that could interfere with signal reception

• Tall buildings and related structures

A prototype of the antenna map made its debut at the 1998 CES. The CEMA Antenna Mapping Program was officially be released at the 1999 CES.

During the course of 1998, CEMA, USSB, and a dozen manufacturers of antennas participated in processes to create standards for testing and labeling antennas, and many other activities. The top 30 markets were made available in December 1998. Remaining markets were available by the end of the first quarter of 1999.

More information about the CEMA antenna mapping project can be found at the CEMA Web site at: http://www.cemacity.org

WHAT IS THE CEMA ANTENNA MAPPING PROGRAM?

The CEMA Antenna Mapping Program is designed to provide retail outlets in 211 U.S. Designated Market Areas (DMAs) with maps showing categories of TV antennas appropriate to receive TV signals in those areas. Selecting the most effective antenna can often be guesswork for sales staff and customers. However, with the mapping program, a customer can select the best antenna type for his or her location in just a few short minutes.

The process works as follows:

Step 1: Customer locates his or her home on the map

Step 2: Determine color zone for customer's home

Step 3: Consult the Antenna Selector Guide to review any special instructions for recommended color zone

Step 4: Match the recommended color zone to the Antenna Selector label on the antenna packaging.

The objective of the mapping program is to provide consumers with a highly accurate guide to select a TV antenna that will work at their homes, and to be able to make that decision in just a few minutes. Field tests have demonstrated that the program is compatible with both analog and digital transmissions.

Unlike older models of antenna maps, the CEMA Antenna Mapping Program is not a series of "concentric circles." Rather, the maps were developed through a highly sophisticated process that takes dozens of variables into account. Transmitter tower placement, topography, tall buildings, water towers, and many other factors that impact signal transmission were calculated into the creation of these maps.

At present, the mapping program is only available in a hard-copy booklet format (see map kit ordering information for details). However, as the program develops, an electronic version will be made available over the Internet. We anticipate a searchable Internet version being available at some point in 1999.

While there is not a perfect solution for every customer when it comes to recommending the most appropriate antenna for each situation, the CEMA Antenna Mapping Program goes a long way to providing the answers consumers want and need.

NOTE: There are some areas not suitable for TV antennas. Reception in downtown areas can be problematic. High-rise buildings can cause a severe ghosting environment. While most of these cases can be solved by a suitable directional antenna, there are a few cases where an antenna may not provide workable reception. Distant outlying areas in very rural locations may not receive a strong enough signal to provide an adequate picture.

SELECTOR GUIDE AND COLOR CATEGORY REVIEW

Antenna maps are broken into seven different color zones. These zones identify that different types of antennas are required in order for a consumer to receive optimal reception. Typically, the closer a consumer lives to the signal tower, the better the reception. The viewer may also be able to use an indoor antenna versus an outdoor. The farther away a consumer lives, the opposite is true. However, there are many variables that impact exactly which antenna a consumer will need.

The antenna selector guide can be read in a clockwise fashion starting with the YELLOW zone. Consumers living in this zone typically live closest to the signal tower. Consumers in the DARK GREEN area typically live slightly farther out and so on. Individuals living in the RED area may have difficulty with ghosting (see Frequently Asked Questions) and need to use any antenna with a RED label.

A copy of the Antenna Selector Guide is included in every antenna map book. This allows quick reference by the salesperson when working with a customer.

OUTDOOR ANTENNA SELECTOR GUIDE

Yellow

Are there any buildings, steeples, towers, or other structures taller than four stories within four blocks of your receive location ?

(See Ghost note below)

If you answered "No," your antenna choices are:

• Small multidirectional antenna

• Large multidirectional antenna installation may work

• Directional antennas

If you answered "Yes," your antenna choices are:

• Medium-size directional antenna

(Outdoor installation is best, but an attic antenna might be suitable see attic note below.)

(See Ghost note below)

Green

Are there any buildings, steeples, towers, or other structures taller than four stories within four blocks of your receive location?

If you answered "No," your antenna choices are:

• Large multidirectional antenna

• Directional antennas

If you answered "Yes," your antenna choices are:

• Medium-size directional antenna

(Outdoor installation is best, but an attic installation may work. See attic note below.)

Light Green

Are there any buildings, steeples, towers, or other structures taller than four stories within four blocks of your receive location ? (See Ghost note below)

If you answered "No," your antenna choices are:

• Roof-mounted, amplified, large multidirectional antenna

• Medium- or large-size directional antenna

If you answered "Yes," your antenna choices are:

• Medium- or large-size directional antenna

Red

Medium- or large-size directional antenna

Blue

Rooftop medium- or large-size directional antenna with antenna-mounted preamplifier

Violet

Rooftop large-size directional antenna with antenna-mounted preamplifier

Pink

Rooftop large-size directional antenna with antenna mounted preamplifier will receive at least one station

White

No TV Service from this city (Check other cities in your area)

- Professional tall tower installation

- Satellite-delivered network stations

- Cable TV

Ghost Note: Many structures can reflect TV signals. This can lead to the TV receiver "seeing" more than one version of the TV signal. When a TV signal arrives at the receiver via more than one path, what is commonly known as *ghosting* results. If the ghosting is caused by a single structure that creates one distinct ghost a large multidirectional antenna may be useful with careful positioning to eliminate the reflected signal. However, the antenna may require different positioning for each channel. Directional antennas are the most ghost resistant antennas since they "see" in only one direction and tend not to see the reflected ghost signal. The farther away from such structures your antenna is, the less likely you are to have a problem, but many factors such as the structure's total surface area and which direction it faces will make a difference on how much effect the structure will have on your TV reception.

Structures that are likely to cause ghosting: Nearby buildings that are taller than your antenna or block your antenna's view in the direction of the TV station's transmitter such as church steeples, apartment or condominium buildings, schools, water towers, industrial buildings, office buildings and warehouses, and large communications (radio, TV) towers, or athletic field lighting towers.

Structures that are usually not a problem for TV reception: Trees and foliage (In some cases foliage can absorb TV signals reducing signal strength, so the best antenna locations will avoid foliage in the direction of the TV station's transmitter) Other homes and residential or commercial wooden buildings that are not taller than your antenna location, street light and utility poles, cellular or PCS poles that are a single pole no higher than treetops, ham radio antennas, or other TV antennas.

Attic Note: An antenna should not be used in an attic unless the antenna selector guide suggests it. Even when that type of antenna will perform well outdoors, reception of TV signals in an attic can be made very difficult due to interference from other electric devices and building construction techniques that may hinder the entrance of the V signal or cause reflections of the signal that lead to ghosting.

INDOOR ANTENNA SELECTOR GUIDE

All indoor antennas will fall into the small multidirectional or large multidirectional antenna category. The CEMA TV Antenna Map can only serve as a guide to where an indoor antenna might work. Even when an antenna will perform well outdoors, reception of TV signals indoors can be made very difficult due to interference from other electric devices and building construction techniques that may hinder the entrance of the TV signal or cause reflections of the signal that lead to ghosting. The Yellow and Green map areas are where indoor antennas are most likely to work well.

Yellow

Are there any buildings, steeples, towers, or other structures taller than four stories within four blocks of your receive location ?

(See Ghost note below)

If you answered "No," your antenna choices are:

• Small multidirectional antenna

• Large multidirectional antenna

If you answered "Yes," your antenna choices are:

• Medium-size outdoor directional antenna (Outdoor installation is best, but an attic installation may work. See attic note below.)

Green

Are there any buildings, steeples, towers, or other structures taller than four stories within four blocks of your receive location ?

(See Ghost note below)

If you answered "No," your antenna choices are:

• Large multidirectional antenna

If you answered "Yes," your antenna choices are:

• Medium-size outdoor directional antenna (Outdoor installation is best, but an attic installation may work. See attic note below.)

Ghost Note: See outdoor antenna selection note.

LIST OF DMAS BY STATE

Following is a list of the Designated Market Areas (DMAs) for the CEMA Antenna Mapping program. Maps for the top 30 markets were made available in December 1998. Maps for remaining markets were to be available by the end of the first quarter of 1999.

It should be noted that the mapping program is believed to provide coverage for 90 to 95 percent of all households in the U.S. There are certain rural areas that do not fall within any specific DMA.

DMAs with Antenna Maps Available in December 1998

Atlanta	Cleveland
Baltimore	Dallas-Ft. Worth
Boston	Denver
Charlotte (2 maps)	Detroit
Chicago	Hartford-New Haven (2 maps)
Cincinnati	Houston

Indianapolis

Los Angeles

Miami-Ft. Lauderdale (2 maps)

Minneapolis-St. Paul

New York

Orlando-Daytona Beach-Melbourne (2 maps)

Philadelphia

Phoenix

Pittsburgh

Portland, Ore.

Raleigh-Durham

Sacramento-Stockton-Modesto

San Diego (2 maps)

San Francisco-Oakland-San Jose

Seattle-Tacoma (2 maps)

St. Louis

Tampa-St. Petersburg-Sarasota

Washington, D.C.

DMAs with Maps Available by First Quarter 1999

Alaska
Anchorage

Fairbanks

Alabama
Anniston

Birmingham

Dothan

Huntsville-Decatur

Mobile-Pensacola (Fla.)

Montgomery

Tuscaloosa

Arkansas
Ft. Smith

Jonesboro

Little Rock-Pine Bluff

Arizona
Tucson-Nogales

Yuma-El Centro

California
Bakersfield

Chico-Redding

Eureka

Fresno-Visalia

Monterey-Salinas

Palm Springs

Santa Barbara- Sanmar-Sanluob

Colorado
Colorado Springs-Pueblo

Grand Junction-Montrose

Delaware
Wilmington

Florida

Ft. Myers-Naples

Gainesville

Jacksonville-Brunswick

Mobile (Ala.)-Pensacola

Panama City

Tallahassee-Thomasville

West Palm Beach-Ft. Pierce

Georgia

Albany

Columbus

Macon

Savannah

Hawaii

Honolulu

Iowa

Cedar Rapids-Waterloo-Dubuque

Davenport-Rock Island (Ill.)-Moline (Ill.)

Des Moines-Ames

Ottumwa-Kirksville

Quincy (Ill.)-Hannibal (Mo.)-Keokuk

Sioux City

Idaho

Boise

Idaho Falls-Pocatello

Twin Falls

Illinois

Champaign-Springfield-Decatur

Davenport (Iowa)-Rock Island-Moline

Monroe-El Dorado

Quincy-Hannibal (Mo.)-Keokuk (Iowa)

Paducah (Ky.)-Cape Girardeau (Mo.)-Mt. Vernon

Peoria-Bloomington

Rockford

Indiana

Evansville

Fort Wayne

Lafayette

Paducah (Ky.)-Cape Girardeau (Mo.)-Mt. Vernon (Ill.)

South Bend-Elkhart

Terre Haute

Kansas

Joplin (Mo.)-Pittsburg

Kansas City

Topeka

Wichita-Hutchinson

Kentucky

Bowling Green

Lexington

Louisville

Paducah-Cape Girardeau (Mo.)-Mt. Vernon (Ill.)

Louisiana

Alexandria

Baton Rouge

Lafayette

Lake Charles

New Orleans

Shreveport

Massachusetts
Springfield-Holyoke

Maryland
Salisbury

Maine
Augusta

Bangor

Portland-Auburn

Presque Isle

Michigan
Alpena

Flint-Saginaw-Bay City

Grand Rapids-Kalamazoo-Battle Creek

Lansing

Marquette

Traverse City-Cadillac

Minnesota
Duluth-Superior (Wis.)

Mankato

Rochester-Mason City (Iowa)-Austin

Missouri
Columbia-Jefferson City

Joplin-Pittsburg (Kan.)

Kansas City

Quincy (Ill.)-Hannibal-Keokuk (Iowa)

Springfield

St. Joseph

Mississippi
Biloxi-Gulfport

Columbus-Tupelo-West Point

Greenwood-Greenville

Hattiesburg-Laurel

Jackson

Meridian

Montana
Billings

Butte

Glendive

Great Falls

Helena

Missoula

Nebraska
Lincoln-Hastings-Kearney

North Platte

Omaha

Nevada
Las Vegas

Reno

New Mexico
Albuquerque-Santa Fe

New York
Albany-Schenectady-Troy

Binghamton

Buffalo

Elmira

Rochester

Syracuse

Utica

Watertown

North Carolina
Greensboro-High Point-Winston-Salem

Greenville-New Bern-Washington

North Dakota
Fargo-Valley City

Minot-Bismarck-Dickinson

Ohio
Columbus

Dayton

Lima

Toledo

Youngstown

Zanesville

Oklahoma
Ada-Ardmore

Oklahoma City

Tulsa

Oregon
Bend

Eugene

Medford-Klamath Falls

Pennsylvania
Erie

Harrisburg-Lancaster-Lebanon-York

Johnstown-Altoona

Wilkes-Barre-Scranton

Rhode Island
Providence-New Bedford

South Carolina
Charleston

Columbia

Florence-Myrtle Beach

Greenville-Spartanburg-Asheville

South Dakota
Rapid City

Sioux Falls-Mitchell

Tennessee
Chattanooga

Jackson

Knoxville

Memphis

Nashville

Tri-Cities (Tenn.-Va.)

Texas
Abilene-Sweetwater

Amarillo

Austin

Beaumont-Port Arthur

Corpus Christi

El Paso

Harlingen-Weslaco-Brownsville

Laredo

Lubbock

Odessa-Midland

San Angelo

San Antonio

Tyler

Victoria

Waco-Temple-Bryan

Wichita Falls-Lawton

Utah
Salt Lake City

Virginia
Charlottesville

Harrisonburg

Norfolk-Portsmouth-Newport News

Richmond-Petersburg

Roanoke-Lynchburg

Tri-Cities (Tenn.-Va.)

Vermont
Burlington-Plattsburgh (N.Y.)

Washington
Spokane

Yakima-Pasco-Richland

West Virginia
Bluefield-Beckley-Oak Hill

Charleston-Huntington

Clarksburg-Weston

Parkersburg

Wheeling-Steubenville (Ohio)

Wisconsin
Green Bay-Appleton

La Crosse-Eau Claire

Madison

Milwaukee

Wausau-Rhinelander

Wyoming
Casper-Riverton

Cheyenne-Scottsbluff (Neb.)-Sterling (Col.)

Index

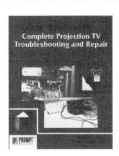

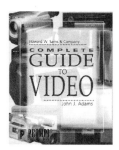

TV Video Systems for the Hobbyist and Technician
L.W. Pena & Brent A. Pena

Covering standard hard-wired cable, large-dish satellite systems, wireless cable, and DSS (Digital Satellite Systems), *TV Video Systems for The Hobbyist & Technician* explains the different systems, how they are installed, their advantages and disadvantages, and how to troubleshoot problems.

Presented in easy-to-understand text and illustrations, it covers installation instructions, home options, apartment options, detecting and repairing problems, and more, guiding you through your TV video project to a successful conclusion.

TV Video Systems for the Hobbyist & Technician covers the following: The Different TV Video Systems, Contracting for TV Video Services, Typical System Components, Apartment Installations, Home Installations, How To Recognize and Overcome Problems, Video Installations–The Right Way, Adding Accessories, and Troubleshooting and Repairing.

Video Technology
124 pages • paperback • 6 x 9"
ISBN: 0-7906-1082-5 • Sams: 61082
$19.95

Complete Guide to Video
John Adams

Complete Guide to Video explains video technology systems in an easy-to-understand language. It outlines the common components of modern audio/video equipment and gives details and features of the newest gadgets. This book was designed to help you with your shopping choices and to answer questions about video technology that the average salesperson may not know. An invaluable reference guide, this book will help you set up the right components for your surroundings and will give you the information you need to get the most out of your video equipment.

Topics include: Home Video Basics, Sights, Sounds & Signals: All About TVs, Antennas, Satellite Dishes & Cable, VCRs, LaserDisc Players & DVD Players, Camcorders, Video & Audio Editing, Game & Internet Consoles, Purchasing Help & Recommendations, Features, and Troubleshooting.

Video Technology
336 pages • paperback • 7-3/8 x 9-1/4"
ISBN: 0-7906-1123-6 • Sams: 61123
$29.95

**To order your copy today or locate your nearest Prompt®
Publications distributor : 1-800-428-7267 or www.hwsams.com**
Prices subject to change.

Complete Guide to Audio
John J. Adams

Practical Acoustics
Stephen Kamichik

Complete Guide to Audio was written for the consumer who wants to know more about sound systems. With comprehensive, simple explanations, it answers questions you may have asked salespeople in the past but were unable to get answers for.

In addition, this book explains some common problems that you may run into while setting up your home entertainment center. The information in the book will help you make successful purchasing decisions and demystify the jungle of wires and connections that come with your audio system.

Topics include: Audio Basics, Sound, Stereo-phonics, Home Theater, Amplifiers and Preamplifiers, Receivers and Surround-Sound, Cassette and CD Decks, DVD, MiniDisc and Phonographs, Speakers, Headphones and Microphones, Computer Sound, Brands and Choices, Hookups and Accessories.

Acoustics is the physics of sound. As simple as that may sound, though, it is actually a complex multidisciplinary science-one that is not taught in most universities and technical schools. A full understanding of acoustics, then, demands that the student grasp the many concepts and practical applications of this fascinating subject.

Practical Acoustics is a two-part in-depth study of this science. Part One contains eleven chapters of theory, such as acoustical waves, decibels, transmission through media, ultrasonics, human ear anatomy, underwater acoustics, microphones and speakers. Each chapter has test questions at the end. Part Two contains eight chapters of projects, such as a speaker enclosure, graphic equalizer, crossover network, ultrasonic auto radar, preamplifier circuits, surround-sound, tone control, and more. Problem solutions and a full glossary of related terms is also included.

Audio
163 pages • paperback • 7-1/4 x 9-3/8"
ISBN: 0-7906-1128-7 • Sams: 61128
$29.95

Audio
259 pages • paperback • 7-3/8 x 9-1/4"
ISBN: 0-7906-1169-4 • Sams 61169
$29.95

To order your copy today or locate your nearest Prompt® Publications distributor : 1-800-428-7267 or www.hwsams.com
Prices subject to change.